Principles and Applications of Fluorescence Spectroscopy

Jihad René Albani
Laboratoire de Biophysique Moléculaire
Université des Sciences et Technologies de Lille
France

Blackwell
Publishing

©2007 by Jihad Rene Albani

Blackwell Science, a Blackwell Publishing company
Editorial offices:
Blackwell Science Ltd, 9600 Garsington Road, Oxford OX4 2DQ, UK
    Tel: +44 (0) 1865 776868
Blackwell Publishing Professional, 2121 State Avenue, Ames, Iowa 50014-8300, USA
    Tel: +1 515 292 0140
Blackwell Science Asia Pty Ltd, 550 Swanston Street, Carlton, Victoria 3053, Australia
    Tel: +61 (0)3 8359 1011

First published 2007

ISBN 978-1-4051-3891-8

Library of Congress Cataloging-in-Publication Data
Albani, Jihad Rene , 1956-
    Principles and applications of fluorescence spectroscopy / Jihad Rene Albani.
        p. ; cm.
    Includes bibliographical references and index.
    ISBN-13: 978-1-4051-3891-8 (pbk.)
    ISBN-10: 1-4051-3891-2 (pbk. : alk. paper) 1. Fluorescence spectroscopy. I. Title.
    [DNLM: 1. Spectrometry, Fluorescence. 2. Biochemistry. QD 96.F56 A326p 2007]
    QP519.9.F56A43 2007
    543'.56–dc22
                            2006100265

A catalogue record for this title is available from the British Library

Set in 10/12 Minion
by Newgen Imaging Systems (P) Ltd, Chennai, India

For further information on Blackwell Publishing, visit our website:
www.blackwellpublishing.com

Contents

Color plate appears between pages 168 and 169

Chapter 1
Absorption Spectroscopy Theory

1.1 Introduction

With reference to absorption spectroscopy, we deal here with photon absorption by electrons distributed within specific orbitals in a population of molecules. Upon absorption, one electron reaches an upper vacant orbital of higher energy. Thus, light absorption would induce the molecule excitation. Transition from ground to excited state is accompanied by a redistribution of an electronic cloud within the molecular orbitals. This condition is implicit for transitions to occur. According to the Franck–Condon principle, electronic transitions are so fast that they occur without any change in nuclei position, that is, nuclei have no time to move during electronic transition. For this reason, electronic transitions are always drawn as vertical lines.

The energy of a pair of atoms as a function of the distance separating them is given by the Morse curve (Figure 1.1). R_e is the equilibrium bond distance. At this distance, the molecule is in its most stable position, and so its energy is called the molecular equilibrium energy, which is expressed as E_0 or E_e. Stretching or compressing the bond induces an energy increase. On the left-hand side of R_e, the two atoms become increasingly closer, inducing repulsion forces. Thus, an energy increase will be observed as a consequence of these repulsion forces. On the right-hand side of R_e, the distance between the two atoms increases, and there will be attraction forces so that an equilibrium distance can be reached. Thus, an energy increase will be observed as result of the attraction forces. In principle, a harmonic oscillation should be obtained, but this is not the case. In fact, beyond a certain distance between the two atoms, the attraction forces will exert no more influence, and attraction energy will reach a plateau. Therefore, the Morse curve is anharmonic.

The energy within a molecule is the sum of several distinct energies:

$$E \text{ (total)} = E \text{ (translation)} + E \text{ (rotation)} + E \text{ (vibration)}$$
$$+ E \text{ (electronic)} + E \text{ (electronic orientation of spin)}$$
$$+ E \text{ (nuclear orientation of spin)} \tag{1.1}$$

The rotational energy concerns molecular rotation around its gravity center, vibrational energy is the result of periodic displacement of atoms of the molecule away from the equilibrium position, and electronic energy is generated by electron movement within the molecular bonds. Rotational levels have lower energy than vibrational ones and thus are lower in energy. Upon photon absorption, electronic, rotational, and vibrational levels

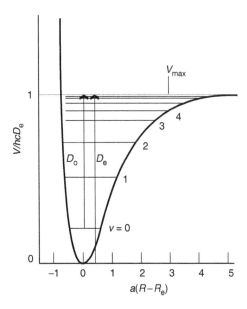

Figure 1.1   Morse curve characterizing the energy of the molecule as a function of the distance R that separates the atoms of a diatomic molecule such as hydrogen. At a distance equal to R_e, which corresponds to point 0, the molecule is in its most stable position, and so its energy is called the molecular equilibrium energy and expressed as E_e. Stretching or compressing the bond yields an increase in energy. The number of bound levels is finite. D_0 is the dissociation energy and D_e the dissociation minimum energy. The horizontal lines correspond to the vibrational levels.

participate in this phenomenon; therefore, energy absorbed by a molecule is equal to the sum of energies absorbed by electronic, vibrational, and rotational energy levels. In fluorescence or phosphorescence, all three levels release energy. Finally, it is important to mention that molecules that absorb photons are called chromophores.

1.2   Characteristics of an Absorption Spectrum

An absorption spectrum is the result of electronic, vibrational, and rotational transitions. The spectrum maximum (the peak) corresponds to the electronic transition line, and the rest of the spectrum is formed by a series of lines that correspond to rotational and vibrational transitions. Therefore, absorption spectra are sensitive to temperature. Raising the temperature increases the rotational and vibrational states of the molecules and induces the broadening of the recorded spectrum.

The profile of the absorption spectrum depends extensively on the relative position of the E_r value, which depends on the different vibrational states. The intensity of the absorption spectrum depends, among others, on the population of molecules reaching the excited state. The more important is this population, the higher the intensity of the corresponding absorption spectrum will be. Therefore, recording absorption spectrum of the same molecule at different temperatures should yield, in principle, an altered or modified absorption spectrum.

A spectrum is characterized by its peak position (the maximum), and the full width at half maximum, which is equal to the difference

$$\delta v = v_2 - v_1 \tag{1.2}$$

v_1 and v_2 correspond to the frequencies that are equal to half the maximal intensity.

If the molecules studied do not display any motions, the spectral distribution will display a Lorenzian-type profile. In this case, the probability of the electronic transition $E_i \rightarrow E_f$ is identical for all molecules that belong to the E_i level.

Thermal motion induces different displacement speeds for the molecules and thus different transition probabilities. These change from one molecule to another and from one population of molecules to another. In this case, the spectral distribution will be Gaussian. The full width at half maximum of a Gaussian spectrum is greater than that of a Lorenzian spectrum.

The spectrophotometer should have a thermostat *in situ* during the experiment and be positioned away from the sun; otherwise, the temperature of the spectrophotometer's inner electronics would increase, possibly introducing important errors in the optical density (OD) measurements. In addition, exposing the spectrophotometer to sunlight can induce continuous fluctuations in the OD, thus making any serious measurements impossible.

Absorption spectrum is the plot of light intensity as a function of wavelength. Figure 1.2 shows the absorption spectra of tryptophan, tyrosine, and phenylalanine in water. A strong band at 210–220 nm and a weaker band at 260–280 nm can be seen.

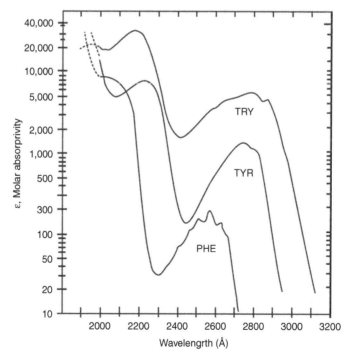

Figure 1.2 Absorption spectra of tryptophan, tyrosine and phenylalanine in water. Source: Wetlaufer, D.B. (1962). *Advances in Protein Chemistry*, **17**, 303–390.

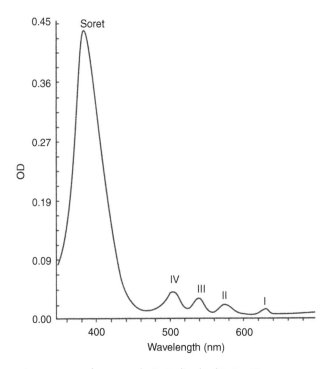

Figure 1.3   Absorption spectrum of protoporphyrin IX dissolved in DMSO.

A molecule can have one, two, or several absorption peaks or bands (Figure 1.3). The band located at the highest wavelength and therefore having the weakest energy is called the first absorption band. In the visible range between 500 and 650 nm, porphyrins display four absorption bands. The intensity ratios between these bands are a function of the nature of lateral chains "carried" by the pyrrolic ring. These bands are the results of electronic transitions of nitrogen atoms of the pyrrolic ring. Metalloporphyrins show two degenerated bands α and β that result, respectively, from the association of bands I and III and bands II and IV.

1.3   Beer–Lambert–Bouguer Law

An absorption spectrum is characterized by two parameters, the maximum position (λ_{max}) and the molar extinction coefficient (ε) calculated in general at λ_{max}. The relation between ε, sample concentration (c), and thickness (l) of the absorbing medium is characterized by the Beer–Lambert–Bouguer law. Since the solution studied is placed in a cuvette, and the monochromatic light beam passes through the cuvette, the thickness of the sample is called the optical path length or simply the path length.

While passing through the sample, light is partly absorbed, and the spectrophotometer will record theoretically nonabsorbed or transmitted light. Plotting the transmittance, which

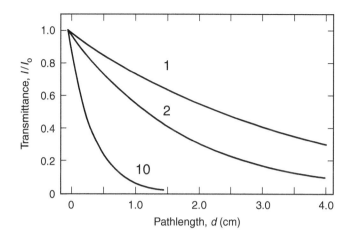

Figure 1.4   Transmittance variation with optical pathlength for three sample concentrations.

is the ratio of transmitted light I_T over incident light I_0:

$$T = I_T/I_0 \tag{1.3}$$

as a function of optical path length and sample concentration yields an exponential decrease (Figure 1.4).

Therefore, T is proportional to exponential $(-cl)$:

$$T \propto e^{-cl} \tag{1.4}$$

Equation (1.4) can also be written as

$$\ln T = \ln(I_T/I_0) \propto -cl \tag{1.5}$$

A proportionality constant k can be introduced, yielding

$$\ln T = \ln(I_T/I_0) = -kcl \tag{1.6}$$

$$-\ln T = \ln(I_0/I_T) = kcl \tag{1.7}$$

Transforming Equation (1.7) to a decimal logarithm, we obtain

$$-\log T = \log(I_0/I_T) = \frac{kcl}{2.3} = \varepsilon cl = OD = A \tag{1.8}$$

Equation (1.8) is the Beer–Lambert–Bouguer law. At each wavelength, we have a precise OD. Since the absorption or OD is equal to a logarithm, it does not have any unity. Concentration c is expressed in molar (M) or mol l^{-1}, the optical path length in centimetres (cm), and thus ε in M^{-1} cm^{-1}.

ε characterizes the absorption of 1 mol l^{-1} of solution. This is called the extinction coefficient because incident light going through the solution is partly absorbed. Thus, the light intensity is "quenched" and is attenuated or inhibited while passing through the solution.

If the concentration is expressed in mg ml^{-1} or in g l^{-1}, the unity of ε will be in mg^{-1} ml·cm^{-1} or g^{-1} l cm^{-1}. Conversion to M^{-1} cm^{-1} is possible by multiplying the value of ε expressed in g^{-1} l cm^{-1} by the chromophore molar mass.

In general, ε is determined at the highest absorption wavelength(s), since molecules absorb most at these wavelengths. However, it is possible to calculate an ε at every wavelength of the absorption spectrum.

In proteins, three amino acids, tryptophan, tyrosine, and phenylalanine, are responsible for UV absorption. ε in proteins is determined at the maximum (278 nm) (Figure 1.2), and thus protein concentrations are calculated by measuring absorbance at this wavelength. Cystine and the ionized sulfhydryl groups of cysteine absorb also in this region but their absorption is weaker than the three aromatic amino acids. Ionization of the sulfhydryl group induces an increase in the absorption and the appearance of a new peak around 240 nm. The imidazole group of histidine absorbs in the 185–220 nm region. Finally, important absorption of the peptide bonds occurs at 190 nm.

When a protein possesses a prosthetic group such as heme, its concentration is usually determined at the absorption wavelengths of the heme. The most important absorption band of heme is called the Soret band and is localized around 408–425 nm. The peak position of the Soret band depends on the heme structure, and in cytochromes, this will depend on whether cytochrome is oxidized or reduced.

Chromophores free in solution and bound to macromolecules do not display identical ε values and absorption peaks. For example, free hemin absorbs at 390 nm. However, in the cytochrome b$_2$ core extracted from the yeast *Hansenula anomala*, the absorption maximum of heme is located at 412 nm with a molar extinction coefficient equal to 120 mM^{-1} cm^{-1} (Albani 1985). In the same way, protoporphyrin IX dissolved in 0.1 N NaOH absorbs at 510 nm, whereas when it is bound to apohemoglobin, it absorbs in the Soret band at around 400 nm.

1.4   Effect of the Environment on Absorption Spectra

The environment here can be the temperature, solvent, chromophore interaction with another molecule, etc.

In general, the interaction between the solvent and the chromophore occurs via electrostatic interactions and hydrogen bonds. In the presence of a highly polar solvent, the dipole–dipole interaction requires a high amount of energy, and thus the position of the absorption peak will be located at low wavelengths. On the contrary, when the dipole–dipole interaction is weak such as when the chromophore is dissolved in a solvent of low polarity, the energy required for absorption is weak, i.e., the position of the absorption band is located at high wavelengths. A blue shift in the absorption band is called the hypsochromic shift, and a red shift is called the bathochromic shift.

When the chromophore binds to proteins, the binding site is generally more hydrophobic than the solution. Therefore, one should expect to observe a shift in the absorption band to the highest wavelength compared to the peak observed when the chromophore is free in solution. The following example of absorption spectroscopy, specifically related to a prosthetic group in proteins, is the vanadate-containing enzyme vanadium chloroperoxidase (VCPO) from the fungus *Curvularia inaequalis*. This enzyme primarily

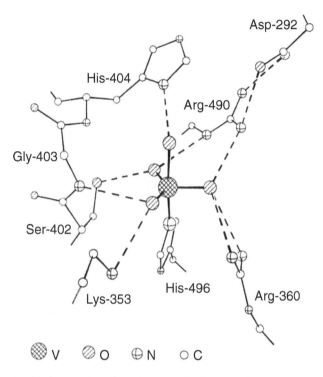

Figure 1.5 Vanadate binding in VCPO from *C. inaequalis*, determined at pH 8.3 (PDB: 1IDQ). Source: Plass, W. (1999). *Angewandte Chemie*, **38**, 909–912. Reprinted with permission from Wiley-Intersciences.

catalyzes halide oxidation to hypohalous acids at the expense of hydrogen peroxide Equation (1.9). *C. inaequalis* is a plant parasite, and VCPO is considered to be involved in the breakdown of the lignin structures of plant leaves (Wever *et al.* 2006). *In vitro*, VCPO also catalyzes sulfoxidation (Andersson *et al.* 1997; Ten Brink *et al.* 1999).

$$H_2O_2 + X^- + H^+ \rightarrow HOX + H_2O \quad X = Cl, Br, \text{ or } I \qquad (1.9)$$

The VCPO active site, displayed in Figure 1.5, shows the binding environment of vanadate (HVO_4^{2-}), as determined by X-ray crystallography (Messerschmidt *et al.* 1997). The cofactor is directly bound to a histidine residue (His496), and the position of the nonprotein vanadate oxygens is stabilized by a hydrogen-bonding network. The resulting structure is a trigonal bipyramid with one apical $N\varepsilon 2$ nitrogen of His496, and the other apical position is considered to be a hydroxide, based on the V—O bond length (1.96 Å). The negative charge on the three equatorial oxygens is stabilized by three positively charged residues, Arg490, Arg360, and Lys353. A minimal catalytic scheme of VCPO is shown in Figure 1.6, also based on the X-ray structure of VCPO crystals soaked in H_2O_2, the first substrate of the enzyme (Messerschmidt *et al.* 1997; Hasan *et al.* 2006).

The UV-VIS spectrum of 100 μM VCPO at pH 8.3 is displayed in Figure 1.7a, which shows the spectra of apo-, holo-, and holo-enzyme after the addition of the first substrate H_2O_2 (peroxo-intermediate) (Renirie *et al.* 2000a). If a halide is added to the peroxo-intermediate, the holo-spectrum is reformed, in line with the scheme shown in Figure 1.6.

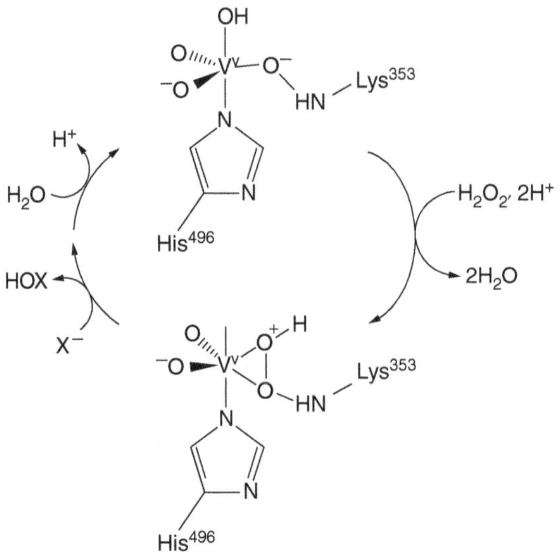

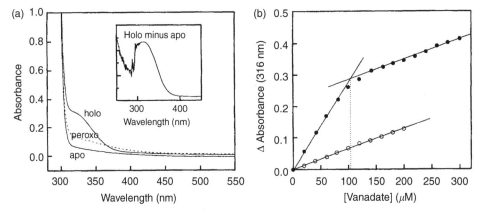

Figure 1.6 Minimal catalytic scheme of VCPO based on crystal structures of the native enzyme and the peroxo-intermediate (Messerschmidt *et al.* 1997). Lys353 is considered to be crucial in assisting heterolytic cleavage of the side-on bound peroxide. EPR and V-EXAFS data suggest that the enzyme remains in the V^V oxidation state throughout the cycle. Source: Hasan, Z., Renirie, R., Kerkman, R., Ruijssenaars, H.J., Hartog, A.F. and Wever, R. (2006). *Journal of Biological Chemistry*, **281**, 9738–9744. Reprinted with permission from Renirie, R., Hemrika, W., Piersma, S.R. and Wever, R. (2000). *Biochemistry*, **39**, 1133–1141. Copyright © 2000 American Chemical Society.

Figure 1.7 (a) UV-VIS absorption spectra of 100 μM vanadium chloroperoxidase at pH 8.3. The lower solid line shows the apo-enzyme, the upper solid line the holo-enzyme after addition of 100 μM vanadate. The dotted line is the peroxo-form of the enzyme after addition of 100 μM H_2O_2. (b) Titration of apo-CPO with vanadate; changes in the optical absorbance at 316 nm (●). Absorbance of free vanadate under the same buffer conditions (○). Reprinted with permission from Renirie, R., Hemrika, W., Piersma, S.R. and Wever, R. (2000). *Biochemistry*, **39**, 1133–1141. Copyright © 2000 American Chemical Society.

The inset shows the difference spectrum of holo minus apo-VCPO, with a maximum at 315 nm, with $\Delta\varepsilon = 2.8$ mM^{-1}. The remaining absorption of the peroxo-intermediate at 315 nm is $\Delta\varepsilon = 0.7$ mM^{-1} cm^{-1}. Titration of apoprotein with vanadate, shown in Figure 1.7b, shows stochiometric binding and a very high affinity of the protein for the cofactor (Renirie *et al.* 2000a).

Figure 1.8 shows VCPO titration with H_2O_2, confirming the high affinity for this substrate as observed in steady-state experiments by Hemrika *et al.* (1999). Figure 1.9 shows that

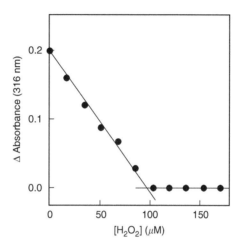

Figure 1.8 Titration of 100 μM VCPO with H_2O_2 at pH 8.3. Source: Renirie, R., Hemrika, W., Piersma, S.R. and Wever, R. (2000). *Biochemistry*, **39**, 1133–1141. Reprinted with permission from The American Chemical Society © 2000.

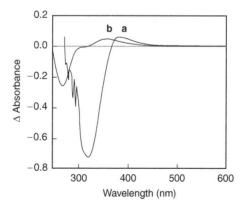

Figure 1.9 Spectrum a: Difference spectrum for 300 μM peroxo-VCPO minus holo-VCPO. Spectrum b: Difference spectrum for 300 μM pervanadate minus vanadate. Source: Renirie, R., Hemrika, W., Piersma, S.R. and Wever, R. (2000). *Biochemistry*, **39**, 1133–1141. Reprinted with permission from The American Chemical Society © 2000.

addition of H_2O_2 also results in a small positive band at 384 nm (a), which is red-shifted as compared to free peroxovanadate (b).

In native VCPO, the position and intensity of cofactor-induced absorption are sensitive to pH change; at pH 5, the band is blue-shifted ($\lambda_{max} \sim 308$ nm), and the intensity is decreased to $\varepsilon = 2.0$ mM^{-1} cm^{-1}. The exact correlation between the above-mentioned spectral features and the known X-ray data are complex, but a further UV-VIS analysis of several mutants of this enzyme showed the importance of several structural elements. A His496Ala mutant showed no optical spectrum (Hemrika *et al.* 1999), whereas crystal structure of this mutant showed the presence of tetrahedral vanadate in the active site (Macedo-Ribeiro *et al.* 1999), indicating that the V—N bond is crucial for the observed spectrum. The effect of other mutations can be seen in Table 1.1.

Figure 1.10 shows the mutation effect of the catalytically important His404 on the spectrum and illustrates how the spectrum can be used to observe changes in the vanadate

Table 1.1 Extinction coefficients of holo minus apo VCPO and mutants and dissociation constants for vanadate determined from absorption spectroscopy at pH 8.3

Enzyme	$\Delta\varepsilon$ (holo−apo)	λ	K_d (vanadate)	Reference
VCPO	2.8 mM^{-1} cm^{-1}	316 nm	<20 μM	Renirie *et al.* (2000a)
H496A	No spectrum observed	—	—	Renirie *et al.* (2000a)
H404A	2.1 mM^{-1} cm^{-1}	316 nm	112 ± 3 μM	Renirie *et al.* (2000b)
R360A	1.5 mM^{-1} cm^{-1}	308 nm	<20 μM	Renirie *et al.* (2000b)
R490A	No spectrum observed	—	—	Renirie *et al.* (2000a)
K353A	No spectrum observed	—	—	Renirie *et al.* (2000a)
D292A	2.8 mM^{-1} cm^{-1}	316 nm	<20 μM	Renirie *et al.* (2000b)

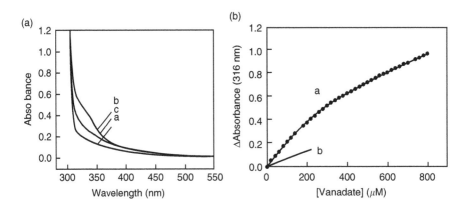

Figure 1.10 UV-VIS absorbance of the H404A mutant of VCPO and the effect of H_2O_2 at pH 8.3. (a) Spectrum a: 200 μM apo-H404A; spectrum b: mixture of holo- and apo-enzyme after addition of 200 μM vanadate; spectrum c: the effect of addition of 200 μM H_2O_2. (b) Spectrum a: titration of 200 μM apo-H404A with 0–800 μM vanadate; the line shown is a fit to the data points for a simple dissociation equilibrium; spectrum b: absorbance of 0–200 μM free vanadate. Source: Renirie, R., Hemrika, W. and Wever, R. (2000). *Journal of Biological Chemistry*, **275**, 11650–11657. Reprinted with permission from The American Society for Biochemistry and Molecular Biology.

dissociation constant. The role of Arg360 on the spectrum is larger, whereas the role of Arg490 and Lys353 is hard to determine, since there is no crystal structure available for the R490A and K353A mutants.

Based on the analogy with extensively studied inorganic imidazole–peroxovanadate complexes (Renirie *et al.* 2000a), it is likely that a charge-transfer transition from the peroxide to the vanadium atom is responsible for the absorption features in the peroxo-form of the enzyme. Interestingly, this suggestion and all the effects of pH, mutation, and H_2O_2 addition were confirmed by extensive DFT calculations, with a dominant His496 $\pi \rightarrow$ V 3d transition in the native VCPO (Borowski *et al.* 2004). In the peroxide form, a peroxide $\pi^* \rightarrow$ V 3d CT transition is most intensive. Calculations predict wavelength shift upon protonation of apical OH. The protonation state of vanadate oxygens is discussed in detail by Pooransingh-Margolis *et al.* (2006).

References

Albani, J. (1985). Fluorescence studies on the interaction between two cytochromes extracted from the yeast *Hansenula anomala. Archives of Biochemistry and Biophysics*, **243**, 292–297.

Andersson, M., Willets, A. and Allenmark, S. (1997). Asymmetric sulfoxidation catalyzed by a vanadium-containing bromoperoxidase. *Journal of Organic Chem*istry, **62**, 8455–8458.

Borowski, T., Szczepanik, W., Chruszcz, M. and Broclawik, E. (2004). First principle calculations for the active centres in vanadium-containing chloroperoxidase and its functional models: geometrical and spectral properties. *International Journal of Quantum Chemistry*, **99**, 864–875.

Hasan, Z., Renirie, R., Kerkman, R., Ruijssenaars, H.J., Hartog, A.F. and Wever, R. (2006). Laboratory-evolved vanadium chloroperoxidase exhibits 100-fold higher halogenating activity at alkaline pH. Catalyic effect from first and second coordination sphere mutations. *Journal of Biological Chemistry*, **281**, 9738–9744.

Hemrika, W., Renirie, R., Macedo-Ribeiro, S., Messerschmidt, A. and Wever, R. (1999). Heterologous expression of the vanadium-containing chloroperoxidase from *Curvularia inaequalis* in *Saccharomyces cerevisiae* and site-directed mutagenesis of the active site residues His[496], Lys[353], Arg[360], Arg[490]. *Journal of Biological Chemistry*, **274**, 23820–23827.

Macedo-Ribeiro, S., Hemrika, W., Renirie, R., Wever, R. and Messerschmidt, A. (1999). X-ray crystal structure of active site mutants of the vanadium-containing chloroperoxidase from the fungus *Curvularia inaequalis. Journal of Biological Inorganic Chemistry*, **4**, 209–219.

Messerschmidt, A., Prade, L. and Wever, R. (1997). Implications for the catalytic mechanism of the vanadium-containing enzyme chloroperoxidase from the fungus *Curvularia inaequalis* by X-ray structures of the native and peroxide form. *Journal of Biological Chemistry*, **378**, 309–315.

Plass, W. (1999). Phosphate and vanadate in biological systems: chemical relatives or more? *Angewandte Chemie*, **38**, 909–912.

Pooransingh-Margolis, N., Renirie, R., Hasan, Z., Wever, R., Vega, A. J. and Polenova, T. (2006). [51]V solid-state magic angle spinning NMR spectroscopy of vanadium chloroperoxidase. *Journal of the American Chemical Society*, **128**, 5190–5208.

Renirie, R., Hemrika, W., Piersma, S. R. and Wever, R. (2000a). Cofactor and substrate binding to vanadium chloroperoxidase determined by UV-VIS spectroscopy and evidence for high affinity for pervanadate. *Biochemistry*, **39**, 1133–1141.

Renirie, R., Hemrika, W. and Wever, R. (2000b). Peroxidase and phosphatase activity of active-site mutants of vanadium chloroperoxidase from the fungus *Curvularia inaequalis. Journal of Biological Chemistry*, **275**, 11650–11657.

Ten Brink, H.B., Holland, H.L., Schoemaker, H.E., van Lingen, H. and Wever, R. (1999). Probing the scope of the sulfoxidation activity of vanadium bromoperoxidase from *Ascophyllum nodosum*. *Tetrahedron: Asymmetry*, **10**, 4563–4572.

Wetlaufer, D.B. (1962). Ultraviolet absorption spectra of proteins and amino acids. *Advances in Protein Chemistry*, **17**, 303–390.

Wever, R., Renirie, R. and Hasan, Z. (2005). Vanadium in biology. In: R. Bruce King (ed.), *Encyclopedia of Inorganic Chemistry* (2nd edn), Wiley, Chichester.

Chapter 2

Determination of the Calcofluor White Molar Extinction Coefficient Value in the Absence and Presence of α_1-Acid Glycoprotein

2.1 Introduction

The purpose of this experiment is to learn how to determine the molar extinction coefficient (ε) value for a ligand and to see how the calculated value is modified upon binding on a macromolecule such as a protein.

Before beginning the experiment, students should be able to describe the functioning of a spectrophotometer, a drawing should accompany the description, and students should be able to demonstrate the Beer–Lambert–Bouguer law.

2.2 Biological Material Used

2.2.1 Calcofluor White

Calcofluor White (Figure 2.1) is a fluorescent probe capable of making hydrogen bonds with β-(1 → 4) and β-(1 → 3) polysaccharides (Rattee and Greur 1974). The fluorophore shows a high affinity for chitin, cellulose, and succinoglycan, forming hydrogen bonds with free hydroxyl groups (Maeda and Ishida 1967).

In the presence of succinoglycan, a polymer of an octasaccharide repeating unit, consisting of galactose, glucose, acetate, succinate, and pyruvate in a ratio of ~1:7:1:1:1 (Åman *et al.* 1981), Calcofluor White fluoresces brightly as the result of its binding to the oligasaccharide (York and Walker 1998).

Calcofluor White is commonly used to study the mechanism by which cellulose and other carbohydrate structures are formed *in vivo* and is also widely used in clinical studies (Andreas *et al.* 2000; Green *et al.* 2000; Srinivasan 2004; Doctor Fungus website, *Candida Endophthalmitis*).

2.2.2 α_1-Acid glycoprotein

α_1-Acid glycoprotein (orosomucoid) is a small acute-phase glycoprotein ($M_r = 41\ 000$) that is negatively charged at physiological pH. It consists of a chain of 183 amino acids (Dente *et al.* 1987), contains 40% carbohydrate by weight, and has up to 16 sialic acid

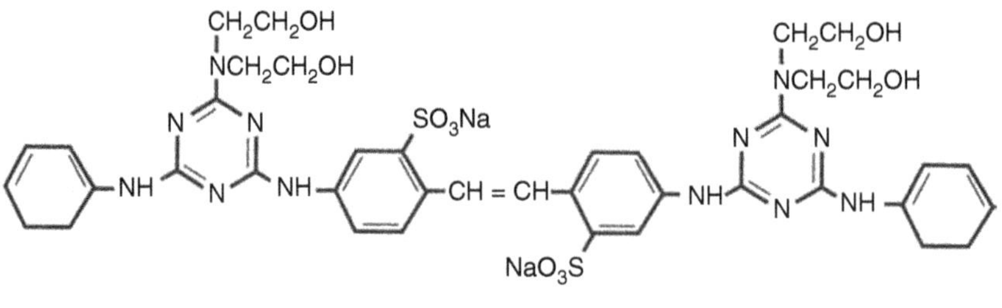

Figure 2.1 Chemical structure of calcofluor white.

20
1 * Arg
Glu-Ile-Pro-Leu-Cys-Ala-Asn-Leu-Val-Pro-Val-Pro-Ile-Thr-Asn-Ala-Thr-Leu-Asp-------Ile-Thr-Gly- Lys -Trp-Phe-
Gln

* 40
Tyr-Ile-Ala-Ser-Ala-Phe-Arg-Asn-Glu-Glu-Tyr-Asn-Lys-Ser-Val-Gln-Glu-Ile-Gln-Ala-Thr-Phe-Phe-Tyr-Phe-Thr-

* 60 Asp Ile * Thr-Thr
Pro-Asn-Lys-Thr-Glu-Asp-Thr-Ile-Phe-Leu-Arg-Glu-Tyr-Gln-Thr-Arg-Gln--------Gln-Cys--------Tyr-Asn------------
Asn Phe Ser-Ser

80 * Ile Val Gln Val 100
Tyr-Leu-Asn-Val-Gln-Arg-Glu-Asn-Gly-Thr---------Ser-Arg-Tyr----------Gly-Gly--------Glu-His------- Ala-His-Leu-
Val Glu Arg Phe

Leu Phe-Gly-Ser-Tyr-Leu-Asp 120
Leu-Ile-Leu-Arg-Asp-Thr-Lys-Thr--------Met-----------------------------------Asp-Glu-Lys-Asn-Trp-Gly- Leu-Ser-
Tyr Leu-Ala-Phe-Asp-Val-Asn

Phe 140
----------Tyr-Ala-Asp-Lys-Pro-Glu-Thr-Thr-Lys-Glu-Gln-Leu-Gly-Glu-Phe-Tyr-Glu-Ala-Leu-Asp-Cys-Leu-
Val

Cys Arg Met 160
----------Ile-Pro---------Ser-Asp-Val---------Tyr-Thr-Asp-Trp-Lys-Lys-Asp-Lys-Cys-Glu-Pro-Leu-Glu-Lys-Gln-His-
Arg Lys Val

180
Glu- Lys-Glu-Arg-Lys-Gln-Glu-Glu-Gly-Glu-Ser.

Figure 2.2 Primary structure of α_1-acid glycoprotein. The five heteropolysaccharide units are linked N-glycosidically to the asparagine residues that are marked with a star. Sources: Schmid, K., Kaufmann, H., Isemura, S. *et al.* (1973). *Biochemistry*, **12**, 2711–2724 and Dente, L., Pizza, M.G., Metspalu, A. and Cortese, R. (1987). *EMBO Journal*, **6**, 2289–2296.

residues (10–14% by weight) (Kute and Westphal 1976). Five heteropolysaccharide groups are linked via an N-glycosidic bond to the asparaginyl residues of the protein (Schmid *et al.* 1973) (Figure 2.2). The protein contains tetra-antennary as well as di- and tri-antennary carbohydrates.

Although the biological function of α_1-acid glycoprotein is still obscure, a number of activities of possible significance have been described, such as the ability to bind the

β-drug adrenergic blocker, propranolol (Sager *et al.* 1979), and certain steroid hormones such as progesterone (Kute and Westphal 1976). Many of these activities have been shown to be dependent on the α_1-acid glycoprotein glycoform (Chiu *et al.* 1977). As the serum concentration of specific α_1-acid glycoprotein glycoforms changes markedly under acute or chronic inflammatory conditions, as well as in pregnancy and tumor growth, a pathophysiological dependence change in the carbohydrate-dependent activities of the protein may occur. Therefore, the relationship between the function of α_1-acid glycoprotein and pathophysiological changes in glycosylation was extensively studied (Mackiewicz and Mackiewicz 1995; Van Dijk *et al.* 1995; Brinkman-Van der Linden *et al.* 1996).

α_1-Acid glycoprotein shows a chemical nature identical to many serum components that have the specific ability of interacting with hormones such as progesterone, to form dissociable complexes. α_1-Acid glycoprotein displays a high affinity for progesterone, but it appears to bind only a small portion of the circulating ligand, and the major part is associated with serum albumin and corticosteroid binding globulin. The interaction between α_1-acid glycoprotein and progesterone is temperature- and pH-dependent (Canguly and Westphal 1968; Kirley *et al.* 1982), and so binding of progesterone to α_1-acid glycoprotein is dependent on the pathophysiological condition. Binding of drugs to circulating plasma proteins such as α_1-acid glycoprotein can decrease the plasma concentration to below the minimal effective concentration, thus inhibiting efficacy and abolishing any therapeutic effect. Recent fluorescence studies helped to describe the α_1-acid glycoprotein tertiary structure: the N-terminal fragment of the protein is in contact with the solvent, and adopts a spatial conformation, a pocket in contact with the buffer and containing one hydrophobic Trp-residue (Albani 2006). Microenvironment of hydrophobic Trp residues of the protein is not compact or rigid. The five carbohydrate units are linked to the pocket and possess a well-defined structure when bound to the α_1-acid glycoprotein (Albani and Plancke 1998, 1999; Albani *et al.* 1999, 2000); their presence in the pocket confers to them their specific structure. Also, fluorescence studies have shown that carbohydrate residues cover most of the interior surface of the pocket (De Ceukeleire and Albani 2002). Therefore, the pocket is formed by two domains, one hydrophilic and the second hydrophobic. Ligands such as progesterone, hemin and 2-*p*-toluidinylnaphthalene-6-sulfonate (TNS) can bind directly to this pocket, since they diffuse from the buffer immediately to the binding site within or at the pocket surface. This binding site is mainly hydrophobic (Albani 2004).

The interaction between Calcofluor White and carbohydrate residues of α_1-acid glycoprotein depends on the secondary structure of the carbohydrate residues, with the fluorescence parameters of Calcofluor being sensitive to this spatial secondary structure.

When dissolved in water, the fluorescence maximum of Calcofluor White is 435–438 nm, while in alcohol such as isobutanol or when bound to human serum albumin, it fluoresces at 415 nm. In the presence of α_1-acid glycoprotein, the fluorescence maximum of Calcofluor shifts toward 439 nm when the fluorophore is at low concentrations and toward 448 nm when it is present at high concentrations. The shift, compared to water and observed in the presence of α_1-acid glycoprotein, is the result of Calcofluor binding on the carbohydrates (40% by weight) of the protein (Albani and Plancke 1998, 1999).

2.3   Experiments

2.3.1   *Absorption spectrum of Calcofluor free in PBS buffer*

1   Plot the baseline from 190 to 450 nm, using PBS buffer in both absorbance cuvettes (buffer vs. buffer). Both cuvettes contain 1 ml of PBS.
2   To the sample cuvette, add with a pipette Pasteur very small quantity of Calcofluor White powder, then mix slowly.
3   Plot absorption spectrum of Calcofluor White from 190 to 450 nm.
4   Prepare a new cuvette with PBS buffer and repeat step 1.
5   Dissolve in the new sample cuvette small quantity of lyophilized α_1-acid glycoprotein. Plot protein absorption spectrum from 200 to 400 nm.
6   Compare the absorption spectra for Calcofluor and α_1-acid glycoprotein; what do you notice? Do you have identical spectra, and can you explain the result? At which wavelength(s) do you have to determine the value of ε of Calcofluor White, and why?

2.3.2   *Determination of ε value of Calcofluor White free in PBS buffer*

Prepare in PBS buffer a stock solution of Calcofluor White equal to 4.24 mg ml^{-1}. To a cuvette containing 1 ml of buffer, add aliquots of 5 μl from the Calcofluor White stock solution. After each addition, mix the solution slowly and measure the optical density at the wavelength you have chosen from the results you obtained in Section 2.3.1. Add at least 10 aliquots, and repeat the experiment twice.

Once you finished the measurements, plot the optical density as a function of Calcofluor White concentration expressed in mg ml^{-1}. What type of plot do you obtain, and why? How do you expect to calculate the value of ε from the plot you obtained? The molecular weight of Calcofluor is equal to 942.7. Can you give the value of ε in M^{-1} cm^{-1}?

2.3.3   *Determination of Calcofluor White ε value in the presence of α_1-acid glycoprotein*

Repeat the experiment described in Section 2.3.2 in the presence of 10 μM of α_1-acid glycoprotein (the ε of the protein at 278 nm in PBS buffer, determined from the same method you are using to calculate the ε of Calcofluor White, is equal to 29.7 mM^{-1} cm^{-1}).

Plot the graphs obtained in Sections 2.3.2 and 2.3.3 on the same paper. What do you notice?

Find in the literature some examples where modification in the value of ε for a chromophore is observed when it is bound to a macromolecule or when the microenvironment is not the same.

Looking at all these results, can you give a common explanation for the modifications observed for all these chromophores?

Find out from the literature how one can estimate the value of ε for a protein, when its amino-acid composition is known. Apply the method to α_1-acid glycoprotein and compare the result to that given in the text.

2.4   Solution

Absorption spectrum of α_1-acid glycoprotein displays two peaks at 225 and 278 nm (Figure 2.3). This feature is characteristic for all proteins. The peak at 278 nm originates from the three aromatic amino acids of the proteins, tyrosine, tryptophan, and phenylalanine. The ε for a protein is generally calculated at 278 nm.

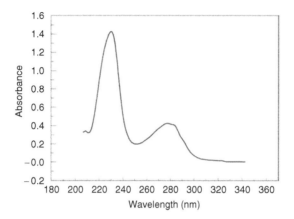

Figure 2.3   Absorption spectrum of 13 μM α_1-acid glycoprotein in PBS buffer, pH 7.

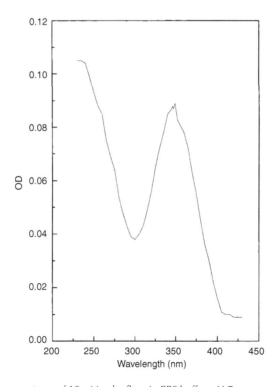

Figure 2.4   Absorption spectrum of 19 μM calcofluor in PBS buffer, pH 7.

Absorption spectrum of Calcofluor displays two peaks, one at 242 nm (clearly observed at high concentrations) and the second at 349 nm (Figure 2.4) We notice that both protein and Calcofluor absorb between 190 and 315 nm. At higher wavelengths, only Calcofluor absorbs, so in order to determine ε for Calcofluor, one should work at wavelengths where Calcofluor only absorbs. The best wavelength will be at the peak equal to 349 nm. In the following experiment, we calculate the value of ε at 352.7 nm.

Plotting the optical density as a function of the Calcofluor concentration yields a linear plot with a slope equal to the product $(l\varepsilon)$ (Figure 2.5). For a cuvette path length l equal to 1 cm, the slope is equal to ε, expressed in mg^{-1} l cm^{-1}. Multiplying the value of ε by the protein's molecular weight yields an ε expressed in mM^{-1} cm^{-1}. The value of ε calculated from the slope is equal to 4.65443 g^{-1} l cm^{-1} or 4387.76 M^{-1} cm^{-1}.

In presence of 10 μM α_1-acid glycoprotein, a value of ε equal to 4.15921 g^{-1} l cm^{-1} or 3920 M^{-1} cm^{-1} is obtained. The value of ε for bound Calcofluor White is around 9% lower than that of free Calcofluor in solution.

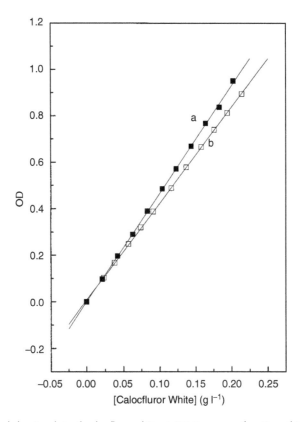

Figure 2.5 Optical density plots of calcofluor white at 352.7 nm as a function of its concentration in (a) the absence and (b) the presence of 10 μM α_1-acid glycoprotein. In the absence of protein, $\varepsilon = $ 4.65443 g^{-1} l cm^{-1} = 4387.76 M^{-1} cm^{-1}. In the presence of the protein, $\varepsilon = 4.15921$ g^{-1} l cm^{-1} = 3920 M^{-1} cm^{-1}. Values of ε are not the same; the value of ε for bound calcofluor white is around 9% lower than that of free calcofluor in solution.

Modification in the value of ε is the result of structural reorganization in the vicinity of Calcofluor White. The presence of a protein alters the electronic distribution within the ligand, inducing a modification of absorption spectrum properties.

We can evaluate ε for a protein by counting the number of tryptophans, tyrosines, and cysteines in the protein sequence and then applying Equation (1.1) (Gill and von Hippel 1989/1990):

$$\varepsilon_{(M^{-1}cm^{-1})} = (5560 \times \text{nb. of Trp}) + (1200 \times \text{nb. of Tyr}) + (60 \times \text{nb. of cys}) \qquad (2.1)$$

α_1-Acid glycoprotein contains three tryptophans, 11 tyrosines, and five cysteines (Figure 2.2). The value of ε determined from Equation (2.1) is 30.410 mM^{-1} cm^{-1}. Thus, the experimental and theoretical values are very similiar.

References

Albani, J.R. (2004) Tertiary structure of human α_1-acid glycoprotein (orosomucoid). Straightforward fluorescence experiments revealing the presence of a binding pocket. *Carbohydrate Research* **339**, 607–612.

Albani, J.R. (2006) Progesterone binding to the tryptophan residues of human α_1-acid glycoprotein. *Carbohydrate Research*, **341**, 2557–2564.

Albani, J.R. and Plancke, Y.D. (1998 and 1999) Interaction between Calcofluor White and carbohydrates of α_1-acid glycoprotein. *Carbohydrate Research* **314**, 169–175 and **318**, 194–200.

Albani, J.R., Sillen, A., Coddeville, B., Plancke, Y.D. and Engelborghs, Y. (1999) Dynamics of carbohydrate residues of α_1-acid glycoprotein (orosomucoid) followed by red-edge excitation spectra and emission anisotropy studies of Calcofluor White. *Carbohydrate Research* **322**, 87–94.

Albani, J.R., Sillen, A., Plancke, Y.D., Coddeville, B. and Engelborghs, Y. (2000) Interaction between carbohydrate residues of α_1-acid glycoprotein (orosomucoid) and saturating concentrations of Calcofluor White. A fluorescence study. *Carbohydrate Research* **327**, 333–340.

Åman, P., McNeil, M., Franzén, L.-E., Darvill, A.G. and Albersheim, P. (1981) Structural elucidation, using h.p.l.c.-m.s. and g.l.c.-m.s., of the acidic polysaccharide secreted by *Rhizobium meliloti* strain 1021[*1,*2]. *Carbohydrate Research* **95**, 263–282.

Andreas S., Heindl, S., Wattky, C., Möller, K. and Rüchel, R. (2000) Diagnosis of pulmonary aspergillosis using optical brighteners. *European Respiratory Journal* **15**, 407.

Brinkman-Van der Linden, E.C., van Ommen, E.C. and van Dijk, W. (1996) Glycosylation of α_1-acid glycoprotein in septic shock: changes in degree of branching and in expression of sialyl Lewis(x) groups. *Glycoconjugate Journal* **13**, 27–31.

Canguly, M. and Westphal, U. (1968, Steroid–protein interactions. XVII. Influence of solvent environment on interaction between human α_1-acid glycoprotein and progesterone. *Journal of Biological Chemistry* **243**, 6130–6139.

Chiu, K.M., Mortensen, R.F., Osmand, A.P. and Gewurz, H. (1977, Interactions of α_1-acid glycoprotein with the immune system. I. Purification and effects upon lymphocyte responsiveness. *Immunology* **32**, 997–1005.

De Ceukeleire, M. and Albani, J.R. (2002) Interaction between carbohydrate residues of α_1-acid glycoprotein (orosomucoid) and progesterone. A fluorescence study. *Carbohydrate Research* **337**, 1405–1410.

Dente, L., Pizza, M.G., Metspalu, A. and Cortese, R. (1987) Structure and expression of the genes coding for human α_1-acid glycoprotein. *EMBO Journal* **6**, 2289–2296.

Doctor Fungus website, *Candida endophthalmitis*. [WWW document]. URL http://216.239.59.104/ search?q=cache:1n8Cng6jaK4J:www.doctorfungus.org/mycoses/human/candida/

Endophthalmitis.htm + clinical + studies, + Calcofluor + white&hl = fr [accessed on 20 March, 2006].

Gill, S.C. and von Hippel, P.H. (1989) Calculation of protein extinction coefficients from amino acid sequence data. *Analytical Biochemistry* **182**, 319–326. Erratum in: *Analytical Biochememistry* 1990, **189**, 283.

Green, L.C., LeBlanc, P.J. and Didier, E.S. (2000) Discrimination between viable and dead encephalitozoon cuniculi (microsporidian) spores by dual staining with Sytox Green and Calcofluor White M2R. *Journal of Clinical Microbiology* **38**, 3811–3814.

Kirley, T.L., Spargue, E.D. and Halsall, H.B. (1982) The binding of spin-labeled propranolol and spin labeled progesterone by orosomucoid. *Biophysical Chemistry* **15**, 209–216.

Kute, T. and Westphal, U. (1976) Steroid–protein interactions. XXXIV. Chemical modification of α_1-acid glycoprotein for characterization of the progesterone binding site. *Biochimica et Biophysica Acta* **420**, 195–213.

Mackiewicz, A. and Mackiewicz, K. (1995) Glycoforms of serum α_1-acid glycoprotein as markers of inflammation and cancer. *Glycoconjugate Journal* **12**, 241–247.

Maeda, H. and Ishida, N. (1967) Specificity of binding of hexopyranosyl polysaccharides with fluorescent brightener. *Journal of Biochemistry* **62**, 276–278.

Rattee, I.D. and Greur, M.M. (1974) *The Physical Chemistry of Dye Absorption.* Academic Press, New York.

Sager, G., Nilsen, O.G. and Jackobsen, S. (1979) Variable binding of propranolol in human serum. *Biochemical Pharmacology* **28**, 905–911.

Schmid, K., Kaufmann, H., Isemura, S. *et al.* (1973) Structure of α_1-acid glycoprotein. The complete amino acid sequence, multiple amino acid substitutions and homology with the immunoglobulins. *Biochemistry* **12**, 2711–2724.

Srinivasan, M. (2004) Fungal keratitis. *Current Opinion in Ophthalmology* **15**, 321–327.

Van Dijk, W., Havenaar, E.C. and Brinkman-Van der Linden, E.C. (1995) α_1-acid glycoprotein (orosomucoid): pathophysiological changes in glycosylation in relation to its function. *Glycoconjugate Journal* **12**, 227–233.

York, G.M. and Walker, G.C. (1998) The *Rhizobium meliloti* ExoK and ExsH glycanases specifically depolymerize nascent succinoglycan chains. *Proceedings of the National Academy of Sciences USA* **95**, 4912–4917.

Chapter 3
Determination of Kinetic Parameters of Lactate Dehydrogenase (LDH)

3.1 Objective of the Experiment

Lactate dehydrogenase (LDH) is an oxidoreductase that catalyzes the conversion of lactate to pyruvate. It consists of four subunits that may be of two different types: M and H ("*muscle*" and "*heart*" formerly known as A and B, respectively). Five different isoenzymes are therefore possible, depending on the subunit composition:

- LDH-1 (H4)
- LDH-2 (H3M)
- LDH-3 (H2M2)
- LDH-4 (HM3)
- LDH-5 (M4)

LDH-1 and LDH-2 are predominant in the heart, while LDH-4 and LDH-5 predominate in skeletal muscle and liver. The molecular weight of all isoenzymes is 140 kDa.

L(+)-Lactate dehydrogenase is specific for L(+)-lactate and does not react with D(−)-lactate. LDH is used in coupled enzyme assays, for example in the determination of ATPase (Penefsky and Bruist 1984), myokinase (Brolin 1983), and pyruvate kinase (Beutler 1971). It may also be used in the determination of lactate (Noll 1984), pyruvate (Lamprecht and Heinz 1984), and various other metabolites.

Thus, the chemical reaction implying LDH is as follows:

$$\text{Pyruvate} + \text{NADH} + \text{H}^+ \underset{2}{\overset{1}{\rightleftharpoons}} \text{lactate} + \text{NAD}^+$$

Students should follow reaction number 1 here. They should follow the disappearance kinetics of NADH with absorption spectrophotometry.

Before entering the laboratory, students should be able to explain a kinetic reaction and to demonstrate how to calculate reaction kinetic constants.

3.2 Absorption Spectrum of NADH

Students should receive a stock solution of NADH with a concentration equal to 5 mg ml^{-1}. Since the molecular weight of NADH is 709, the concentration of the stock solution is

$$\frac{5}{709} = 7.05 \times 10^{-3}\,\text{M} = 7.05\,\text{mM}$$

The absorption spectrum of NADH should be obtained by adding 10 μl of the stock solution to 1.2 ml of phosphate buffer (0.1 M, pH 7.5). The concentration of NADH stock solution should be calculated at 340 nm using an ε equal to 6200 $\text{M}^{-1}\,\text{cm}^{-1}$.

3.3 Absorption Spectrum of LDH

Students should take 0.2 ml of the commercially solution (around 2 mg) and centrifuge it to eliminate the ammonium sulfate solution. The pellet obtained should be dissolved in 2 ml of phosphate buffer (0.1 M, pH 7.5). The concentration of the solution should be 1 mg ml^{-1}. Plot the absorption spectrum of LDH and measure its concentration at 280 nm using an extinction coefficient equal to 1 (mg/ml)$^{-1}\,\text{cm}^{-1}$.

3.4 Enzymatic Activity of LDH

Students should prepare three stock solutions of pyruvate: 150, 30, and 6 mM. 16.5 mg of pyruvate (MW = 110) dissolved in 1 ml of phosphate buffer, pH 7.5, yields a solution of 150 mM. Dilution of this solution five times yields a stock solution of 30 mM, and its dilution 25 times yields a stock solution equal to 6 mM.

In a test tube, add: 170 μl of the NADH stock solution, 100 μl of the 30 mM pyruvate stock solution, the appropriate volume of enzyme and then make up the quantity to 3 ml with phosphate buffer. Mix slowly then measure the variation of absorbance at 340 nm for 2 min. Calculate from the data obtained the specific enzyme activity.

The volumes of LDH added to each 3 ml solution prepared should be as follows: 10, 20, 30, 40, and 50 μl.

3.5 Kinetic Parameters

Prepare a stock solution of NADH equal to 11 mM. In a test tube, add 100 μl of the NADH solution, different volumes of pyruvate stock solution, and 40 μl of LDH stock solution diluted 100 times, and then make up the quantity to 3 ml with phosphate buffer. After mixing the solution, measure the optical density at 340 nm after 2 min reaction. The pyruvate volumes to be added are shown in Table 3.1.

3.6 Data and Results

The absorbance peak of NADH is at 340 nm. The optical density measured should be 0.37, which yields a stock solution concentration of 7.16 mM. The optical density of LDH at 280 nm is 1.1, which yields a stock solution equal to 1.1 mg ml^{-1}.

Table 3.1 Volumes of added pyruvate and final concentrations in the test tubes

Vol. of 30 mM pyruvate solution	900 μl	500	400	300	200	100		50						
Vol. of 6 mM pyruvate solution							500	400	300	250	200	150	100	50
$[S_0]$ mM	9	5	4	3	2	1	0.8	0.6	0.5	0.4	0.3	0.2	0.1

3.6.1 Determination of enzyme activity

Enzyme activity can be obtained by keeping the substrate concentration constant and by varying the enzyme concentration. The following differences in absorption should be obtained after 2 min reaction:

Enzyme volume (μl)	10	20	30	40	50
$\Delta A/2$ min		0.12	0.25	0.35	0.48	0.58

Plotting $\Delta A/2$ min as a function of enzyme volume gives the plot shown in Figure 3.1. For 1 min, the variation in absorption is

$$0.48 \ \Delta A/2 \ \text{min} = 0.24 \ \Delta A \ \text{min}^{-1}$$

This variation could be expressed in terms of the concentration of NADH:

$$0.24 \ \Delta A \ \text{min}^{-1} 6200 \ \text{M}^{-1} \ \text{cm}^{-1} = 38.7 \ \mu\text{M of NADH transformed min}^{-1}$$

The concentration of enzyme used during 1 min of kinetic:

$$(0.01 \ \text{mg ml}^{-1} \times 40 \ \mu\text{l})/3000 = 0.000134 \ \text{mg ml}^{-1} = 0.000134 \ \text{g l}^{-1}$$

1 mol weighs 130 000, and so the concentration of enzyme in the absorbance cuvette is

$$0.000134 \ \text{g l}^{-1}/130 \ 000 = 1 \text{nM}$$

The enzyme activity is

$$38.7 \times 10^{-6} \ \text{mol of NADH min}^{-1} \ \text{l}^{-1}/10^{-9} \ \text{mol of enzyme l}^{-1}$$

$$= 38 \ 700 \ \text{mol of NADH min}^{-1} \ (\text{mol of enzyme})^{-1}$$

Since we have 1 mol of NADH for 1 mol of pyruvate, the enzymatic activity can be expressed as

$$38 \ 700 \ \text{mol of pyruvate min}^{-1} \ (\text{mol of enzyme})^{-1}$$

3.6.2 Determination of kinetic parameters

The substrate concentrations during the experiment and corresponding initial velocities are listed in Table 3.2.

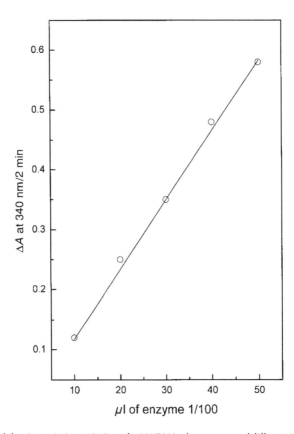

Figure 3.1   Optical density variation at 340 nm for NADH in the presence of different LDH concentrations.

Table 3.2   Values of pyruvate concentrations in the test tubes, and corresponding velocities

$[S_0]$ mM	9	5	4	3	2	1	0.8	0.6	0.5	0.4	0.3	0.2	0.1
$V_0 \Delta A/2$ min	0.385	0.48	0.49	0.545	0.55	0.52	0.465	0.425	0.41	0.355	0.355	0.25	0.15
						0.49			0.38				

Plotting $\Delta A/2$ min as a function of $[S_0]$ gives the graph displayed in Figure 3.2. At excess concentrations of pyruvate, a velocity decrease can be observed. Kinetic parameters of the reaction can be obtained by plotting the reverse data, i.e., ($1/v_0$ vs. $1/[S_0]$).

The inverse of the maximum velocity V_{max} is obtained at the intersection of the plot with the y-axis (Figure 3.3). This yields a value of V_{max} equal to

$$1/1.43 = 0.7 \ \Delta A/2 \text{ min}$$

Expressed in 1 min and in terms of the NADH concentration, V_{max} is

$$0.7/(6200 \text{ M}^{-1} \text{ cm}^{-1} \times 2) = 5.6 \times 10^{-5} \text{ M min}^{-1}$$

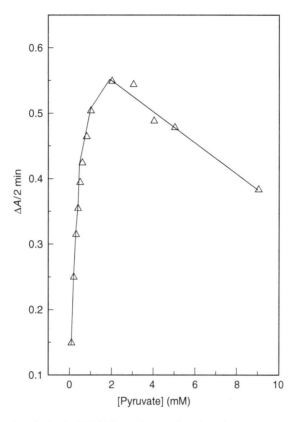

Figure 3.2 Variation in velocity for NADH formation as a function of pyruvate concentration.

The inverse of the Michaelis constant K_m is obtained at the intersection of the plot with the x-axis. This yields a value of

$$1/2.7 = 0.37 \text{ mM}$$

The catalytic constant k_{cat} is equal to the ratio $V_{max}/[E_0]$. Thus, we have to calculate the enzyme concentration in the cuvette. The concentration of the enzyme stock used for the experiment is 0.011 mg ml^{-1}. We put 40 μl of 10^{-2} mg ml^{-1} enzyme in 3 ml.

The enzyme concentration in the experimental solution is

$$(0.011 \text{ mg ml}^{-1} \times 40 \text{ } \mu\text{l})/3000 \text{ } \mu\text{l} = 1.47 \times 10^{-4} \text{ mg ml}^{-1} = 1.47 \times 10^{-4} \text{ g l}^{-1}$$

1 mol weighs 130 000 g. The enzyme concentration should be expressed as the number of moles dissolved in 1 l of solution:

$$[E_0] = \frac{1.47 \times 10^{-4} \text{ g l}^{-1}}{130\,000 \text{ g}} = 1.13 \times 10^{-9} \text{ mol l}^{-1}$$

$$k_{cat} = V_m/[E_0] = 5.6 \times 10^{-5} \text{ M min}^{-1}/10^{-9} \text{ mol} = 56\,000 \text{ min}^{-1}$$

$$= 933 \text{ s}^{-1} \text{ by the tetramer of LDH.}$$

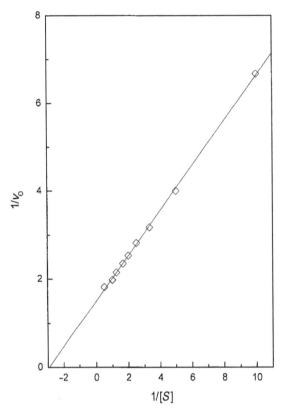

Figure 3.3 Inverse plot of the velocity of disappearance of NADH as a function of pyruvate concentration.

3.7 Introduction to Kinetics and the Michaelis–Menten Equation

3.7.1 Definitions

a. The enzyme active site is composed of the amino acids that interact with the substrate.
b. The active site of an enzyme contains two sites: substrate binding and catalytic sites.
c. The catalytic site acts on the substrate to transform it to one or many products.
d. Enzymes are not modified structurally before and after the reaction, act in small quantities, and do not modify the equilibrium of a reversible reaction. They accelerate the reaction rate without disturbing the final equilibrium (Figure 3.4).

3.7.2 Reaction rates

The reaction rate is measured by quantifying the product formed under the enzyme action or the substrate transformed in one unit time.

The molecular activity of an enzyme is the number of substrate molecules transformed in 1 min by one enzyme molecule under optimal conditions of temperature and pH.

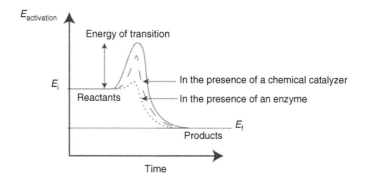

Figure 3.4 Activation energy of a biological reaction with time in the absence and presence of a chemical catalyzer and an enzyme.

3.7.2.1 Zero-order kinetics

This concerns a reaction where the quantity of substrate transformed per time unit is constantly independent of its concentration.

$$\frac{-dS}{dt} = k \tag{3.1}$$

where S is the substrate concentration at a certain time t. The minus sign indicates that the substrate concentration is decreasing:

$$-dS = k\,dt \tag{3.2}$$

$$\int -dS = \int k\,dt \tag{3.3}$$

$$-S = kt + cte \tag{3.4}$$

At $t = 0$, $cte = -S_0$, which is the substrate concentration at the beginning of the reaction. Equation (3.4) can be written as

$$S = -kt + S_0 \tag{3.5}$$

Plotting S as a function of time t yields a linear graph with a slope equal to $-k$ and an intercept at the y-axis equal to S_0.

The same analysis can be done by following the variation in product concentration:

$$\frac{dP}{dt} = k \tag{3.6}$$

$$\int dP = \int k\,dt \tag{3.7}$$

$$P = kt + cte \tag{3.8}$$

At $t = 0$, $cte = 0$.

$$P = kt \tag{3.9}$$

The units of k are M s^{-1}.

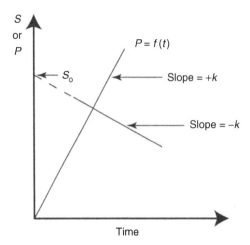

Figure 3.5 Variation of substrate and product concentrations with time for a zero-order kinetics reaction.

Plotting P as a function of time yields a line that begins at zero with a slope of k (see Figure 3.5).

3.7.2.2 First-order kinetics

This concerns a reaction where the quantity of substrate transformed per unit time is proportional to the quantity of the substrate present in solution at the time of the measurement.

The rate of disappearance of the substrate decreases constantly.

$$A \xrightarrow{k} B \tag{3.10}$$

$$v = k[A] = -\frac{d[A]}{dt} = \frac{d[B]}{dt} \tag{3.11}$$

$k = v/[A]$, which means that the dimensions of k are: $M\,s^{-1}/M = s^{-1}$.

From the definition of the first-order kinetics, one can write:

$$-\frac{dS}{dt} = kS \tag{3.12}$$

$$-\frac{dS}{S} = k\,dt \tag{3.13}$$

$$\int -(dS/S) = \int k\,dt \tag{3.14}$$

$$-\ln S = kt + cte \tag{3.15}$$

At $t = 0$, $cte = -\ln S_0$. Thus, Equation (3.15) can be written as

$$-\ln S = kt - \ln S_0 \tag{3.16}$$

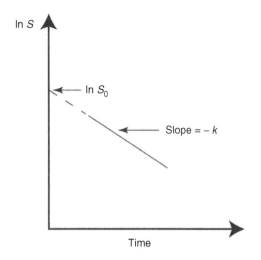

Figure 3.6 Determination of the kinetic constant of a first-order kinetics reaction.

or

$$\ln S = -kt + \ln S_0 \qquad (3.17)$$

Plotting $\ln S$ as a function of t yields a straight line with a slope equal to $-k$ and a y-intercept equal to $\ln S_0$ (Figure 3.6).

Equation (3.17) can be modified and written as:

$$\ln \frac{S}{S_0} = -kt \qquad (3.18)$$

$$\frac{S}{S_0} - e^{-kt} \qquad (3.19)$$

$$S = S_0 e^{-kt} \qquad (3.20)$$

Thus, S decreases exponentially with time. When $S = S_0/2$, we obtain

$$\ln \frac{1}{2} = -kt_{1/2} \qquad (3.21)$$

or

$$\ln 2 = kt_{1/2} \qquad (3.22)$$

$$t_{1/2} = 0.693/k \qquad (3.23)$$

$t_{1/2}$ is the half-life of the kinetic reaction.

3.7.2.3 Substrate-to-product transformation in the presence of enzyme

The role of the enzyme in these types of reactions is to accelerate the rate of product formation. However, the substrate possesses a binding site on the enzyme, and so in order to transform the substrate to a product, an enzyme–substrate complex is formed. The formation of the product follows the following scheme:

$$E + S \underset{k_{-1}}{\overset{k_1}{\rightleftharpoons}} ES \overset{k_2}{\longrightarrow} E + P \tag{3.24}$$

where ES is the intermediate complex (the Michaelis–Menten complex), and k_2 is the calalytic constant. Product formation occurs from the intermediary complex ES. Thus, the velocity, v, of the product formation follows a first-order kinetic rule:

$$v = k_2[ES] \tag{3.25}$$

Quantifying $[ES]$ is difficult, so it should be replaced with other known values.
  The rate of formation of ES is

$$\frac{d[ES]}{dt} = k_1[S_f][E_f] \tag{3.26}$$

where $[S_f]$ and $[E_f]$ are the concentrations of free substrate and enzyme, respectively. We can replace $[E_f]$ by $([E_0] - [ES])$. Equation (3.26) can be written as

$$\frac{d[ES]}{dt} = k_1[S_f]([E_0] - [ES]) \tag{3.27}$$

The rate of disappearance of $[ES]$ is

$$-\frac{d[ES]}{dt} = k_2[ES] + k_{-1}[ES] \tag{3.28}$$

The formation rate of ES is equal to its disappearance rate:

$$k_1[S_f]([E_0] - [ES]) = k_2[ES] + k_{-1}[ES] \tag{3.29}$$

$$k_1[S_f][E_0] - k_1[S_f][ES] = k_2[ES] + k_{-1}[ES] \tag{3.30}$$

$$k_1[S_f][E_0] = k_1[S_f][ES] + k_2[ES] + k_{-1}[ES] \tag{3.31}$$

$$k_1[S_f][E_0] = [ES](k_1[S_f] + k_{-1} + k_2) \tag{3.32}$$

$$[ES] = \frac{k_1[S_f][E_0]}{k_1[S_f] + k_2 + k_{-1}} \tag{3.33}$$

Dividing both the numerator and denominator by k_1 yields

$$[ES] = \frac{[S_f][E_0]}{[S_f] + (k_2 + k_{-1})/k_1} \tag{3.34}$$

$(k_2 + k_{-1})/k_1$ can be replaced by a constant K_m, called the Michaelis–Menten complex:

$$[ES] = \frac{[S_f][E_0]}{[S_f] + K_m} \tag{3.35}$$

The velocity, v, of the product formation is

$$v = k_2[ES] \tag{3.36}$$

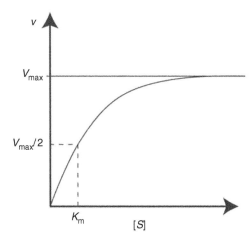

Figure 3.7   Schematic representation of the Michaelis–Menten equation.

Thus, combining Equations (3.35) and (3.36) yields:

$$v = \frac{k_2[S_f][E_0]}{[S_f] + K_m} \tag{3.37}$$

For simplification, one could consider that the concentration of bound substrate is too small compared to the added substrate $[S]$, and so in Equation (3.37), we can replace $[S_f]$ by $[S]$.

$$v = \frac{k_2[S][E_0]}{[S_f] + K_m} \tag{3.38}$$

Equation (3.38) is known as the Michaelis–Menten equation. We have three constant parameters, $k_2, [E_0]$, and K_m. The two variables are $[S]$ and v. When $[S] = 0, v = 0$. When $[S] \to \infty$, $v \to k_2[E_0] = V_{max}$. At infinite or high substrate concentrations, v tends to reach a maximum value V_{max} (Figure 3.7).

   If the value of V_{max} is not reached, plotting $1/v$ vs. $1/[S]$ yields a straight line which intercepts the x-axis at $1/K_m$ and the y-axis at $1/V_{max}$.

3.7.2.4   Units of the measured parameters and constants

Equation (3.34) shows clearly that the units of K_m are those of the substrate, in moles or M.

$$v = k_2[ES] \Rightarrow k_2 = \frac{v}{[ES]} = \frac{M\ s^{-1}}{M} = s^{-1} \tag{3.39}$$

$$K_m = \frac{k_2 + k_{-1}}{k_1} \tag{3.40}$$

The units of k_2 and k_{-1} are identical, in s^{-1}.

$$k_1 = \frac{k_2 + k_{-1}}{K_m} = M^{-1}\ s^{-1} \tag{3.41}$$

The dimensions of V_{max} are identical to those of v, i.e., M s^{-1}.

3.7.2.5 Relationship between K_m and the dissociation constant K_d of the complex ES

$$K_d = \frac{[S_f][E_f]}{[ES]} = \frac{[S_f]([E_0] - [ES])}{[ES]} \tag{3.42}$$

$$\Rightarrow \quad K_d[ES] = [S_f][E_0] - [ES][S_f] \tag{3.43}$$

$$\Rightarrow \quad K_d[ES] + [ES][S_f] = [S_f][E_0] \tag{3.44}$$

$$\Rightarrow \quad [ES][S_f + K_d] = [S_f][E_0] \tag{3.45}$$

$$\Rightarrow \quad [ES] = \frac{[S_f][E_0]}{K_d + [S_f]} \tag{3.46}$$

With the Michaelis constant, we have the following equation:

$$[ES] = \frac{[S_f][E_0]}{K_m + [S_f]} \tag{3.47}$$

Equations (3.46) and (3.47) are equal, and so K_m is apparently equal to K_d of ES. However, under which conditions does this apply?

$$K_m = \frac{k_2 + k_{-1}}{k_1} \tag{3.48}$$

If $k_{-1} >> k_2$, then

$$E + S \underset{k_{-1}}{\overset{k_1}{\rightleftarrows}} ES$$

is favored compared to product formation. In this case, we have

$$K_m = \frac{k_{-1}}{k_1} \tag{3.49}$$

Also, in this case, we can write

$$k_1[S_f][E_f] = k_{-1}[ES] \tag{3.50}$$

$$\Rightarrow \quad K_d = \frac{[S_f][E_f]}{[ES]} = \frac{k_{-1}}{k_1} \tag{3.51}$$

Thus, K_m and K_d are equal only if $k_{-1} \gg k_2$.

References

Beutler, E. (1971) *Red Cell Metabolism, A Manual of Biochemical Methods*, pp. 56–68. Grune & Stratton New York.

Brolin, S.E. (1983) Adenylate kinase (myokinase): UV-method. In: H.U. Bergmeyer (ed.), *Methods of Enzymatic Analysis* (3rd edn), Vol. 3, pp. 540–545. Verlag Chemie, Weinheim, Federal Republic of Germany.

Lamprecht, W. and Heinz, F. (1984) Pyruvate. In: H.U. Bergmeyer (ed.), *Methods of Enzymatic Analysis* (3rd edn), Vol. 6, pp. 555–561. Verlag Chemie, Weinheim.

Noll, F. (1984) L(+)-lactate. In: H.U. Bergmeyer (ed.), *Methods of Enzymatic Analysis* (3rd edn), Vol. 6, pp. 582–528. Verlag Chemie, Weinheim.

Penefsky, H.S. and Bruist, M.F. (1984) Adenosinetriphosphatases. In: H.U. Bergmeyer (ed.), *Methods of Enzymatic Analysis* (3rd edn), Vol. 4, pp. 324–328. Verlag Chemie, Weinham.

Chapter 4
Hydrolysis of *p*-Nitrophenyl-*β*-D-Galactoside with *β*-Galactosidase from *E. coli*

4.1 Introduction

β-Galactosidase is a tetrameric enzyme that consists of identical subunits with a molecular mass of 135 000. Amino-acid analysis indicates approximately 1170 residues per subunit. Each monomer is composed of five compact domains and a further 50 additional residues at the N-terminal end. Within the cell, *β*-galactosidase cleaves lactose to form glucose and galactose. The latter is exchanged with lactose via a lactose–galactose antiport system. The composition and structure of the cell wall change continuously during plant development. Thus, the cell wall is dynamic, not static. Plant cell walls consist of cellulose microfibrils coated by xyloglucans and embedded in a complex matrix of pectic polysaccharides (Talbott and Ray 1992; Carpita and Gibeaut 1993). Plant development involves a coordinated series of biochemical processes that, among other things, result in the biosynthesis and degradation of cell-wall components. Enzymes such as *β*-galactosidase and *α*-arabinosidase play a role in the cross-linking of pectins and cell-wall proteins by catalyzing the formation of phenolic-coupling activity. The enzymes hydrolyze corresponding phospho-*p*-nitrophenylderivatives (*α*-L-arabinofuranoside and *β*-D-galactopyranoside, respectively) into *p*-nitrophenol (PNP) (Figure 4.1) that absorbs at 410 nm. Thus, it is possible to follow the evolution of plant development by quantifying the PNP formed, as shown in a study performed by Stolle-Smits *et al.* (1999) by following the absorption of PNP at 420 nm $(4.8 \times 10^3 \text{ M}^{-1} \text{ cm}^{-1})$. For more information on the properties of *β*-galactosidase, students can look to the following link:

http://www.mpbio.com/product_info.php?cPath=491_1_12&products_id=150039&depth=nested&keywords=beta%20galactosidase

The purpose of the present experiments is to characterize kinetics parameters of *p*-nitrophenyl-*β*-D-galactoside hydrolysis with *β*-galactosidase. Students will find out how to use absorption spectroscopy to study enzymatic properties of an enzyme (see also Murata *et al.* 2003). Before entering the lab, students should be able to explain the Beer–Lambert–Bouguer law and the basics of enzymology.

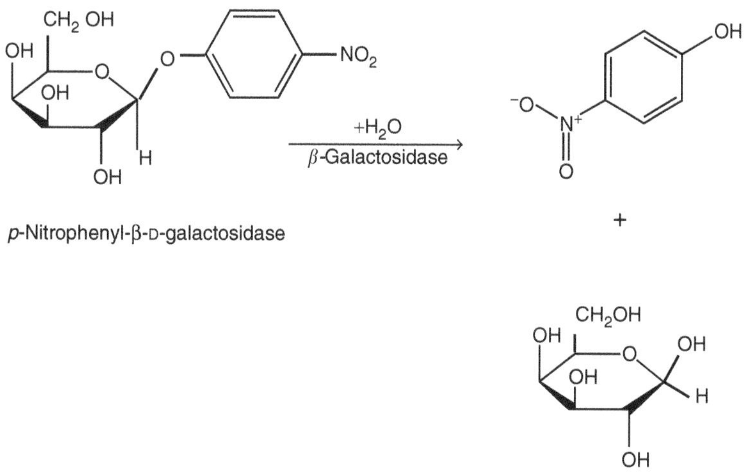

Figure 4.1 Reaction describing the hydrolysis of *p*-nitrophenyl-*β*-D-galactosidase into *p*-nitrophenol and *β*-D-galactose.

4.2 Solutions to be Prepared

- *β*-Galactosidase is purchased as a suspension in 60% saturation of ammonium sulfate solution. The solution given to students corresponds to 1.5 ml of the purchased protein dissolved in 500 ml of buffer pH 7. The diluted solution is stable at the maximum 1 week at 4°C.
- 1 mM heteroside solution: Dissolve 0.03 g of *p*-nitrophenyl-*β*-D-galactosidase in 100 ml of distilled water containing 10 mg of sodium azide.
- 10 mM heteroside solution: Dissolve 0.3 g of *p*-nitrophenyl-*β*-D-galactosidase in 100 ml of distilled water containing 10 mg of sodium azide.
- Stock solution of paranitrophenol at 10^{-6} mol ml^{-1}: Dissolve 139.11 mg in 1 l of distilled water containing 1 drop of concentrated HCL solution.
- 1 M sodium carbonate solution: Dissolve 106 g in 1 l of distilled water.
- Buffers from pH 2 to 12: Buffers can be phosphate or Tris. Add 0.1 g of sodium azide to each 1 l of buffer prepared.

All the experiments are performed in test tubes of internal diameter equal to 1.5 cm.

4.3 First-day Experiments

During the first day, students will discover the absorption properties of PNP and will determine the optimal temperature and pH of *β*-galactosidase.

4.3.1 Absorption spectrum of PNP

Plot an absorption spectrum from 340 to 480 nm of 0.1 ml of 1 mM PNP solution mixed with 4.9 ml of Na_2CO_3 vs 0.1 ml of water mixed to 4.9 ml of Na_2CO_3. Calculate $\varepsilon_{\lambda\,max}$.

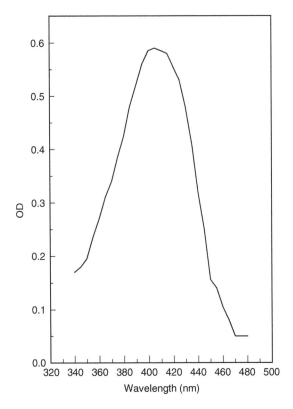

Figure 4.2   Absorption spectrum of paranitrophenol in basic medium.

The absorption spectrum of PNP (Figure 4.2) displays a maximum around 405 nm. The optical density (OD) recorded at this wavelength is slightly higher than that recorded at 410 nm. The value of ε at 405 nm and at any wavelength can be obtained from the Beer–Lambert–Bouguer law:

$$\varepsilon_{405\ nm} = \frac{OD_{405\ nm}}{cxl} = \frac{0.59}{(10^{-3}\ M/50) \times 1.5\ cm} = 19\ 666\ M^{-1}\ cm^{-1}$$

4.3.2   Absorption of PNP as a function of pH

Measure the OD at the absorption maximum (in our case, 405 nm) of 0.5 ml of 1 mM PNP mixed to 4.5 ml buffer vs. 0.5 ml water mixed with 4.9 ml buffer. Experiments should be performed with buffers at different pHs (from 2 to 10). Then, plot the measured ODs as a function of pH. The plot is displayed in Figure 4.3.

4.3.2.1   Conclusion

PNP absorbs only in the deprotonated form, and so in order to observe its formation after the enzymatic reaction, we need to increase the pH in the test tube. This can be achieved by

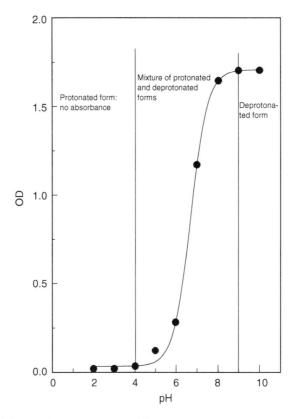

Figure 4.3   Optical density of PNP at 405 nm at different pHs.

adding Na_2CO_3 to the enzyme–substrate solution. Addition of Na_2CO_3 helps to stop the enzymatic reaction and allows the OD of the formed PNP to be read.

4.3.3   Internal calibration of PNP

The purpose of this experiment is to determine up to which PNP concentration, the Beer–Lambert–Bouguer law is still linear. In fact, this law cannot be applied when the optical saturation of the photomultiplier is reached and/or when the chromophore aggregates. Each spectrophotometer cannot count more than a precise number of photons. When this number is reached, the OD given by the spectrophotometer will be the highest, and the photomultiplier is saturated. Beyond the "saturation concentration," the spectrophotometer will always give the same OD that corresponds to that of saturation.

Also, Beer–Lambert–Bouguer law can no longer be applied because of the chromophore itself. At high concentrations, aggregation could occur. In this case, we are going to have two values of ε, corresponding to low and high concentrations. Plotting OD as a function of macromolecule concentration yields two plots with two different ε values.

Prepare two series of test tubes. In the first series, add 0, 0.5, 1, 1.5, and 2 ml of the 0.1 mM PNP solution. In the second series, add 0.2, 0.5, 1, 1.5, and 2 ml of the 1 mM PNP solution. Make up the solution to 2 ml with distilled water, add 3 ml of Na_2CO_3, and mix well.

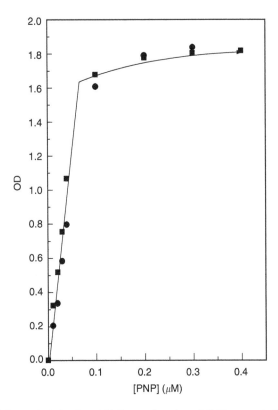

Figure 4.4 Plots of the optical density of PNP as a function of PNP concentration measured at 405 (squares) and 430 nm (circles).

Measure the ODs at 405 nm and 430 nm vs. the blank solution where PNP is replaced by water.

Plot the OD as a function of PNP concentration in the tube at the two wavelengths and determine for each wavelength the maximum PNP concentration at which the Beer–Lambert–Bouguer law is still linear.

The graph (Figure 4.4) shows that the linearity of the Beer–Lambert–Bouguer law can be applied up to ODs equal to around 1.6. This yields a PNP concentration of 0.0675 and 0.1 mM at 405 and 430 nm, respectively. The molar extinction coefficient at 430 nm is lower than that at 405, and it can be calculated from the data of Figure 4.2. In fact, the value of ε at 430 nm is 16 000 M^{-1} cm^{-1}.

Plots obtained show that above 0.07 mM, OD continues to increase, indicating that PNP is probably aggregating. The aggregate does not have the same absorbance properties as the monomer.

4.3.3.1   Conclusion

In order to perform qualitative and quantitative studies with β-galactosidase, we should work in experimental conditions such that only the monomer of PNP is detected. Also, one should work below the detection limit of the spectrophotometer; otherwise we are going to read the same OD at all high concentrations of PNP present in solution.

One should work at an enzyme concentration that yields sufficient PNP to be observed with the spectrophotometer but not in excess, so as to avoid aggregation and/or OD saturation.

Figure 4.4 indicates also that working at 430 nm allows the detection of higher PNP concentrations than when the experiment is performed at 405 nm. However, in both cases, the ODs measured by the end of the hydrolysis should be lower than 1.6.

4.3.4   Determination of β-galactosidase optimal pH

Experiments should be performed using pH buffers going from 2 to 10. Add to a test tube: 0.5 ml of 1 mM heteroside, 1 ml of buffer, and 2 ml of enzyme. Then, mix with the vortex and allow to incubate for 3 min. Finally, add 3 ml of Na_2CO_3.

In the blank tube, add the 3 ml of Na_2CO_3 before adding the enzyme. Plot the OD at 405 nm as a function of pH.

The result (Figure 4.5) shows that the optimum pH of the enzyme is 7. Thus, at this pH, the activity of β-galactosidase is the most important. Therefore, in the following

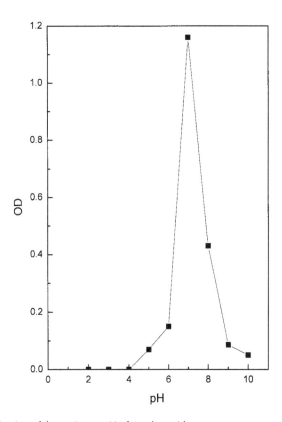

Figure 4.5   Determination of the optimum pH of β-galactosidase.

experiments, pH 7 buffer will be used. Adding Na_2CO_3 induces a high pH, which denatures the enzyme and stops hydrolysis. At the same time, it will be possible to read the OD of the PNP formed.

4.3.5 Determination of β-galactosidase optimal temperature

Enzyme activity is temperature-dependent. Parameters such as temperature and pH affect enzyme structure. Enzyme activity is most important in its native form, modifying the native structure with the temperature, or the pH decreases enzyme activity. In the native form, an enzyme is most active at its optimal pH and temperature. Physiological properties of macromolecules are modified with temperature variation. Molecules will lose their physiological properties only when they reach a temperature that denatures their molecular structure. However, one should be cautious when studying folding and/or unfolding of proteins, since the different regions of the proteins do not behave similarly with temperature. In fact, denaturation of an enzyme active site with temperature could be more difficult than other regions of the protein.

The optimal pH of β-galactosidase was determined at ambient temperature, 20°C. However, this temperature is not necessarily the optimal one. Hydrolysis of the heteroside should be performed at the following temperatures: 0°C, 10°C, 20°C, 30°C, 40°C, 50°C, 60°C, and 100°C.

In a test tube, mix 0.5 ml of 1 mM heteroside and 1 ml of pH 7 buffer. Then, keep the test tube for 5 min at the temperature of the experiment. Then, add 0.5 ml of the enzyme solution without removing the test tube from its position, mix the whole solution slowly, and allow the hydrolysis to continue for 5 min. Then, add 3 ml of Na_2CO_3. A blank tube is prepared by adding Na_2CO_3 before the enzyme. Read the OD at 405 nm at room temperature, and then plot it as a function of temperature.

The data (not shown) indicate that the optimal temperature is 35°C. However, in order to decrease the enzyme activity slightly, the hydrolysis experiments can be performed at 25°C.

4.4 Second-day Experiments

Students will learn how to measure enzyme kinetics and how to determine protein concentration using Bradford's method (Bradford, 1976).

4.4.1 Kinetics of p-nitrophenyl-β-D-galactoside hydrolysis with β-galactosidase

Add to 12 test tubes: 0.5 ml of 1 mM heteroside and 1.9 ml of pH 7 buffer. Mix with the vortex and maintain the tubes for 5 min at 25°C. Then, add 0.1 ml of enzyme solution, mix with the vortex, and put the tubes back at 25°C. After 1, 2, 3, 4, 5, 6, 8, 10, 20, 30, 45 and 60 min, add 3 ml of Na_2CO_3. Read the OD at 405 nm vs a blank tube where Na_2CO_3 was added before the enzyme. Then, plot the OD as a function of time (Figure 4.6). Check that the measured ODs correspond to the linear part of the plot of Figure 4.4.

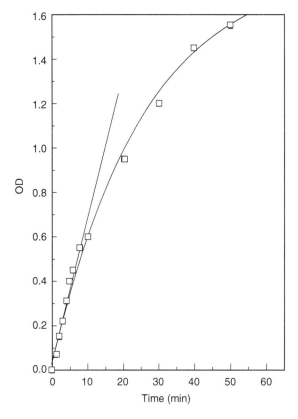

Figure 4.6 Kinetics of *p*-nitrophenyl-β-D-galactoside hydrolysis with β-galactosidase.

First, we can see that the measured ODs fit within the linear part of the plot of Figure 4.4. The ODs correspond to PNP resulting from *p*-nitrophenyl-β-D-galactoside hydrolysis. From the slope of the linear part of the kinetics, we can calculate the hydrolysis velocity, i.e., the PNP concentration formed per minute.

At 15.5 and 5 min, the corresponding ODs are 1.2 and 0.4. Thus, the slope of the linear part is equal to

$$\text{Slope} = \frac{OD_1 - OD_2}{t_1 - t_2} = \frac{1.2 - 0.4}{15.5 - 5.0} = \frac{0.8}{10.5} = 7.62 \times 10^{-2} \text{ min}^{-1} \qquad (4.1)$$

Knowing the path length and the molar extinction coefficient, we can calculate the equivalent concentration of PNP observed:

$$[\text{PNP}] = \frac{7.62 \times 10^{-2}}{19\,666 \text{ M}^{-1} \text{ cm}^{-1} \times 1.5 \text{ cm}} = 2.58 \times 10^{-6} \text{ mol l}^{-1} \text{ min}^{-1} \qquad (4.2)$$

The ODs shown in Figure 4.6, and so the measured concentrations of PNP are those obtained in 5.5 ml of final solution after the addition of Na$_2$CO$_3$. The real ODs and the real concentration of PNP formed are higher since hydrolysis was performed in 2.5 ml.

Thus, the real concentration of PNP obtained in the linear part is

$$[PNP]_{hydrolysis} = \frac{2.58 \times 10^{-6} \times 5.5}{2.5} = 5.68 \times 10^{-6} \text{ mol l}^{-1} \text{ min}^{-1} \qquad (4.3)$$

In order to calculate enzyme activity, one should divide PNP concentration obtained per minute by the enzyme concentration in the test tube. In order to determine the enzyme concentration in the test tubes, we will apply the Bradford method (Bradford, 1976).

4.4.2 Determination of the β-galactosidase concentration in the test tube

In Bradford's method (Bradford, 1976), the OD of Coomassie Blue, a dye that absorbs at 465 nm when it is free in solution and at 595 nm when it is bound to a protein, is followed. Up to certain protein concentrations, the absorption of Coomassie Blue bound to the protein increases linearly. Determination of the protein concentration is obtained using standard proteins such as bovine serum albumin (BSA).

Transfer an appropriate volume of stock solution of BSA (250 μg ml^{-1}) into disposable cuvettes to produce the following BSA concentrations: 0, 2, 4, 8, 12, 16, and 20 μg ml^{-1}. BSA concentrations in the cuvettes are calculated for a final volume of 2 ml.

Add distilled water so that the volume in the cuvettes is 1.6 ml, and finally make up the volume to 2 ml by adding 0.4 ml of Coomassie Blue. Mix slowly, then measure the OD at 595 nm. Plot the OD as a function of BSA concentration (standard plot).

In other disposable cuvettes, prepare different solutions of β-galactosidase from the stock solution. The following enzyme dilutions should be carried out for a final volume of 2 ml: 1/10, 1/15, 1/20, 1/25, 1/30, 1/40, and 1/50 .

Make up the volume with distilled water up to 1.6 ml. Then, add 0.4 ml of Coomassie Blue. Mix slowly, and measure the OD at 595 nm. Concentrations of the different enzyme dilutions can be determined using the standard plot. The concentration of β-galactosidase stock solution is obtained by multiplying the enzyme concentration in the cuvette by the dilution factor.

Figure 4.7 clearly indicates that up to 20 μM BSA, recorded ODs are linearly proportional to serum albumin concentrations. Thus, we can use the standard curve to determine the β-galactosidase concentration:

OD	0.12	0.18	0.2	0.21	0.27	0.37	0.47
Dilution	1/50	1/40	1/30	1/25	1/20	1/15	1/10
[β-Galactosidase]$_{cuvette}$ (μg ml^{-1})	4.8	7.3	8.1	8.6	11	15.1	19.1
[β-Galactosidase]$_{stock}$ (μg ml^{-1})	240	292	243	215	220	226	190

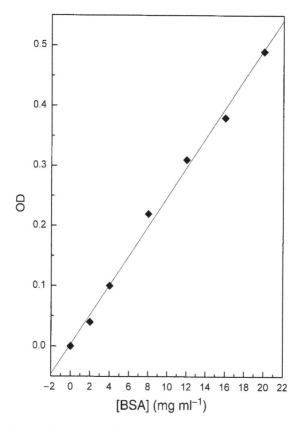

Figure 4.7 Optical density of Coomassie Blue as a function of serum albumin concentrations.

The mean value of the β-galactosidase stock solution is 232 μg ml^{-1}. The β-galactosidase concentration used in the kinetics experiment is (0.1 ml × 232 μg ml^{-1}): 2.5 ml = 9.28 μg ml^{-1}. The activity of β-galactosidase is

$$\frac{[\text{PNP}]_{\text{hydrolysis}}}{[\text{Enzyme}]} = \frac{5.68 \times 10^{-6}\text{mol l}^{-1}\text{ min}^{-1}}{9.28 \times 10^{-3}\text{g l}^{-1}} = 0.612 \times 10^{-3}$$

$$= 6.12 \times 10^{-4} \text{ mol of PNP min}^{-1}\text{ g}^{-1}\text{enzyme} \qquad (4.4)$$

4.4.2.1 Suggested experiment

We suggest here that students determine β-galactosidase activity by performing hydrolysis for a constant time and at different enzyme concentrations.

4.5 Third-day Experiments

Students should be able to describe the theory of Michaelis–Menten before entering the lab.

4.5.1 Determination of K_m and V_{max} of β-galactosidase

The instructor should give the students two stock solutions of *p*-nitrophenyl-β-D-galactoside (1 and 10 mM). Students should then prepare the following heteroside concentrations in a final volume of 2 ml: 0.1, 0.25, 0.5, 0.75, 1, 2.5, 5, 7.5, and 10 mM.

From each of these new stock solutions, prepare the following test tubes. First, add 0.5 ml of heteroside solution, then 1.2 ml of pH 7 buffer, and finally 0.3 ml of enzyme solution. Mix slowly and then keep the hydrolysis reaction going for 3 min at 20°C. Then, add 3 ml of Na_2CO_3. Measure the ODs vs. a blank where Na_2CO_3 is added before the enzyme. From the plot, OD vs. [*p*-nitrophenyl-β-galactoside] (Figure 4.8), calculate V_{max} and K_m. From now on, plot the graphs using the real ODs, i.e., those you should measure in the volume of hydolysis solution, and not those you measured after addition of 3 ml of Na_2CO_3.

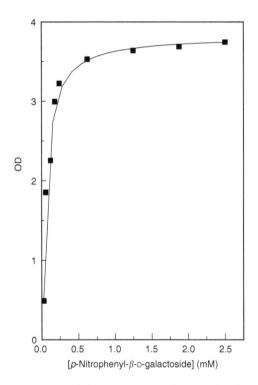

Figure 4.8 Optical density variation with the concentration of [*p*-nitrophenyl-β-galactoside]. The optical densities correspond to 2 ml of hydrolysis solution.

The value of K_m corresponds to the concentration of substrate where velocity is equal to half of V_{max}. K_m is 0.425×10^{-3} mol $l^{-1} = 4.25 \times 10^{-4}$ mol l^{-1}. The Michaelis constant given in the literature for the PNP hydrolysis with β-D-galactosidase is 4.45×10^{-4} M. In

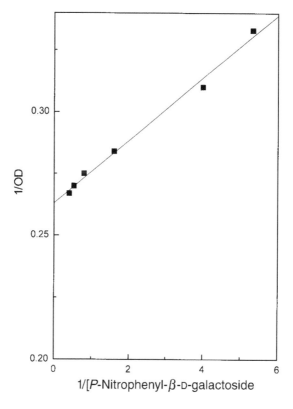

Figure 4.9 Double reciprocal plot of PNP optical density vs. [*p*-nitrophenyl-*β*-D-galactoside] at high heteroside concentrations.

order to be sure that at the highest concentration (2.5 mM) of heteroside, the measured OD corresponds or is very close to the maximum value, we have plotted the inverse graph $1/v$ vs. $1/[p$-nitrophenyl-β-galactoside] (Figure 4.9). The value of V_{max} obtained upon extrapolation on the y-axis is 3.8, thus clearly indicating that the maximum velocity has been reached.

V_{max} can be expressed as the concentration of formed PNP:

$$V_{max} = \frac{OD}{\varepsilon x l} = \frac{3.75}{19\,666 \times 1.5} = 1.27 \times 10^{-4} \text{ mol of PNP } l^{-1}\text{min}^{-1} \tag{4.5}$$

4.5.1.1 Suggested experiment

Students can repeat the last experiment in the presence of two different inhibitor concentrations to determine its effect on the hydrolysis rate.

4.5.2 Inhibiton of hydrolysis kinetics of p-nitrophenyl-β-D-galactoside

The instructor should give students three stock solutions of inhibitor: 5, 10, and 30 mM. Students are going to prepare two series of six test tubes, containing each of the following inhibitor concentrations: 0.5, 1, 1.5, 2, 2.5, and 3 mM. The final volume is 2 ml. To one series, add 0.5 ml of 1 mM heteroside stock solution, and to the second series, add 0.8 ml from the same heteroside stock solution. Thus, the heteroside concentrations used for the hydrolysis are 0.25 and 0.4 mM, respectively. Then, complete with pH 7 buffer up to 1.5 ml, mix, and add 0.5 ml of stock enzyme. Mix slowly, and allow the hydrolysis to continue for 3 min at 20°C. Then, add 3 ml of Na_2CO_3. Measure the ODs vs. blank where Na_2CO_3 is added before the enzyme. Plot the graph $1/v$ vs. $[I]$. Derive the mathematical equation that allows you to describe the type of inhibition, then determine the nature of the inhibition and calculate the value of the inhibition constant K_i and the kinetic constants of the enzyme.

Plotting $1/v$ vs. $[I]$ at the two heteroside concentrations yields the graphs of Figure 4.10. As we can see from the high values of $1/v$, the reaction rate was expressed here in PNP concentration per minute.

The plots obtained correspond to a competitive inhibition characterized by the following equation:

$$\frac{1}{v_i} = \frac{1}{V_{max}} + \frac{K_m}{V_{max}S} + \frac{K_m I}{K_i V_{max}S} \tag{4.6}$$

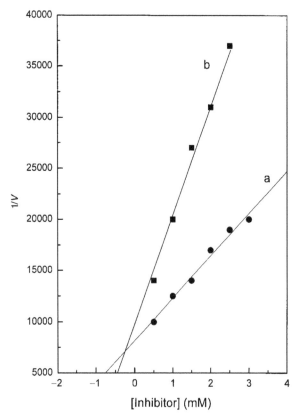

Figure 4.10 Plot of the velocity inverse as a function of inhibitor concentration, in presence of (a) 0.25 and (b) 0.4 mM *p*-nitrophenyl-*β*-D-galactoside.

where v_i is the reaction rate in presence of inhibitor I, K_i is the inhibitor constant corresponding to the dissociation constant of the enzyme–inhibitor complex, and S is the substrate concentration.

Plotting $1/v_i$ as a function of I, yields a linear plot of slope equal to

$$\frac{K_m}{K_i V_{max} S}$$

and a y-intercept equal to

$$\frac{1}{V_{max}} + \frac{K_m}{V_{max} S}$$

When S increases, the slope decreases.

When

$$\frac{1}{v_i} = \frac{1}{V_{max}}, \; I = -K_i$$

Thus, plotting $1/v_i$ as a function of I for different (two to three) substrate concentrations yields linear plots that intercept together at a point having as coordinates ($-K_i$ and $1/V_{max}$). From the graph, we obtain $1/V_{max}$ equal to 7000, i.e., a V_{max} equal to 1.42×10^{-4} mol min^{-1} and K_i equal to 0.28 mM.

From the y-intercept value, we can calculate K_m value. The y-intercepts are equal to 8166 and 10 000 for plot (a) ([heteroside] = 0.4 mM) and plot (b) ([heteroside] = 0.25 mM), respectively. Thus, we have 0.663×10^{-4} mol l^{-1} and 1.065×10^{-4} mol l^{-1} for plots (a) and (b), respectively.

4.6 Fourth-day Experiments

4.6.1 Effect of guanidine chloride concentration on β-galactosidase activity

Protein denaturation induces a loss of its secondary and tertiary structures. Since the tertiary structure is responsible for proteins' physiological properties, its loss induces total protein inactivation. Denaturation occurs in different ways: thermal, pH, urea, guanidine salts (mainly guanidine chloride), organic solvents, and detergents. Denaturation does not occur instantaneously and depends not only on the nature of denaturing factor or agent but also on the macromolecule concentration and structure. The most effective denaturant is guanidine chloride. Total loss of the molecules structure occurs at 6 M GndCl. The salt is a strong electrolyte, which means that it interacts with the protein via electrostatic charges. Protein charges that maintain the folded structure are neutralized by electrostatic interaction with guanidine chloride, thereby inducing an unfolded protein. Also, interaction of guanidine chloride occurs with amino acids of a protein inner core, thereby inducing an irreversible unfolded state.

In the following experiment, we will study β-galactosidase denaturation by guanidine chloride following OD of PNP at 405 nm. Enzyme denaturation induces a loss in its activity, which will be observed by the decrease in the OD.

Students should be given a solution of 6 M guanidine. Then, they should prepare the following guanidine chloride stocks: 0, 0.05, 0.1, 0.15, 0.20, 0.25, 0.30, 0.35, 0.40, 0.45, and 0.50 M. Then, in 11 test tubes each containing 4 ml of stock enzyme, add 1 ml of the guanidine solution stocks. Mix and leave the mixture for 15 min. Then, add 0.25 ml of the mixture to test tubes containing 0.5 ml of 10 mM heteroside and 1.75 ml of pH 7 buffer. Mix and allow the reaction to continue for 25 min at $37°C$. Then, add 3 ml of Na_2CO_3 and measure the ODs at 405 nm vs. a blank where Na_2CO_3 was added before the enzyme. Finally, plot the corrected OD for the volume (2.5 ml) of hydrolysis and for 1 min as a function of guanidine chloride concentration in the test tubes. The same experiment should be done at $25°C$ also.

Derive the mathematical equation that allows you to calculate the equilibrium constant K_{eq} at each guanidine concentration and plot $K_{eq} = f$ ([guanidine]. Finally, determine $\Delta G^{°\prime}$ at each guanidine concentration and plot $\Delta G^{°\prime} = f$ ([guanidine] and $\Delta G^{°\prime} = f$ ([K_{eq}]). Calculate from the last graph the value of K_{eq}.

Do the graphs obtained correspond to the theoretical significance of $\Delta G^{°\prime}$ and K_{eq}?

Do you obtain the same information from both experiments, at 25 and $37°C$?

4.6.2   OD variation with guanidine chloride

Figure 4.11 displays the OD variation of PNP formed in the presence of different guanidine chloride concentrations. We notice that a loss of total enzyme activity occurs at 8 mM guanidine chloride. The enzyme concentration was small, which explains the total loss of activity at concentrations of guanidine chloride much lower than 6 M. The fact that OD reaches zero indicates that with 8 mM guanidine, all β-galactosidase present in the test tube is in a denatured form. In other words, denaturation of the enzyme by guanidine is total and not partial. The denatured structure of β-galactosidase is no longer active.

The hyperbolic shape of the curve in Figure 4.11 allows a dissociation constant to be obtained for the guanidine-β-galactosidase complex equal to 2.2 mM.

4.6.3   Mathematical derivation of K_{eq}

The absorption values of native and denatured β-galactosidase in the presence of a certain concentration of guanidine are denoted as A_n and A_d, respectively. In the presence of guanidine concentrations where equilibrium exists between the two protein states, the recorded absorbance $A_{(obs)}$ is

$$A_{(obs)} = A_n + A_d \tag{4.7}$$

$$A_{(obs)} = \varepsilon_N C_N l + \varepsilon_D C_D l = \varepsilon_N (C_o - C_D) l + \varepsilon_D C_D l \tag{4.8}$$

where ε_N and C_N are the molar extinction coefficient and concentration of the native form, ε_D and C_D are the molar extinction coefficient and concentration of the denatured form, and C_o is the total enzyme concentration in the solution.

$$A_{(obs)} = \varepsilon_N C_o l - \varepsilon_N C_D l + \varepsilon_D C_D l \tag{4.9}$$

$$= A_{(N)} + C_D l (\varepsilon_D - \varepsilon_N) \tag{4.10}$$

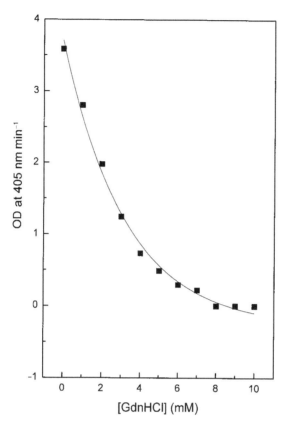

Figure 4.11 Activity variation of β-galactosidase ($18.5\ \mu g\ ml^{-1}$) as a function of guanidine concentration. The experiment was performed at $37°C$.

$A_{(N)}$ and $A_{(D)}$ denote enzyme absorptions when it is completely native or denatured.

$$A_{(obs)} = A_N X_D C_o l (\varepsilon_D - \varepsilon_N) \tag{4.11}$$

where X_D is the fraction of the denatured form.

$$A_{(obs)} = A_{(N)} + X_D C_o l \varepsilon_D - X_D C_o l \varepsilon_N \tag{4.12}$$

$$= A_{(N)} + X_D A_{(D)} - X_D A_{(N)} \tag{4.13}$$

$$= A_{(N)} + X_D [A_{(D)} - A_{(N)}] \tag{4.14}$$

$$A_{(obs)} - A_{(N)} = X_D [A_{(D)} - A_{(N)}] \tag{4.15}$$

$$X_D = \frac{A_{obs} - A_N}{A_D - A_N} \tag{4.16}$$

$$K_{eq} = \frac{C_D}{C_N} = \frac{X_D C_o}{X_N C_o} = \frac{X_D}{X_N} \tag{4.17}$$

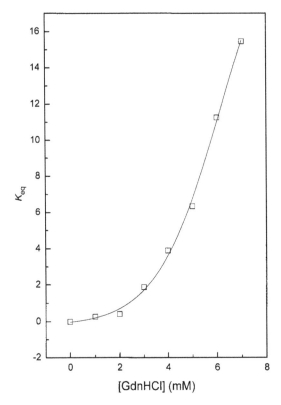

Figure 4.12   Plot of K_{eq} as a function of [guanidine]. At high guanidine concentrations, β-galactosidase is completely denatured, yielding an infinite value of K_{eq}.

$$X_N = 1 - X_D = \frac{A_D - A_N - A_{obs} + A_N}{A_D - A_N} = \frac{A_D - A_{obs}}{A_D - A_N} \tag{4.18}$$

$$K_{eq} = \frac{A_{obs} - A_N}{A_D - A_{obs}} \tag{4.19}$$

Thus, it is possible to calculate the equilibrium constant at each guanidine chloride concentration using Equation (4.19). $K_{eq} = f$ ([guanidine]) yields the graph shown in Figure 4.12.

How do we interpret the results obtained for the equilibrium constant variation? Let us consider the following chemical reaction:

$$aA + bB \underset{2}{\overset{1}{\rightleftharpoons}} cC + dD \tag{4.20}$$

When the system is in equilibrium, we can write:

$$K_{eq} = \frac{[C]^c[D]^d}{[A]^a[B]^b} \tag{4.21}$$

$$\text{If } K_{eq} > \frac{[C]^c[D]^d}{[A]^a[B]^b} \Rightarrow K_{eq}[A]^a[B]^b > [C]^c[D]^d \tag{4.22}$$

In this case, reactants are in excess, and the reaction will occur in direction 1.

$$\text{If } K_{eq} < \frac{[C]^c[D]^d}{[A]^a[B]^b} \Rightarrow K_{eq}[A]^a[B]^b < [C]^c[D]^d \tag{4.23}$$

In this case, products are in excess, and the reaction will occur in direction 2. The purpose of this displacement, whether in direction 1 or 2, is to reach equilibrium.

If $K_{eq} < 1 \Rightarrow$ [reactants] > [products]

If $K_{eq} > 1 \Rightarrow$ [reactants] < [products]

In our experiments, although equilibrium between native and denatured forms of β-galactosidase exists at all guanidine chloride concentrations, at high concentrations the enzyme will be in one form, denatured, and thus there will be no kinetic equilibrium.

4.6.4 Definition of the standard Gibbs free energy, $\Delta G^{\circ\prime}$

The Phase Rule describes the possible number of degrees of freedom in a (closed) system at equilibrium, in terms of the number of separate phases and the number of chemical constituents in the system. It was deduced from thermodynamic principles by J. W. Gibbs in the 1870s. (http://jwgibbs.cchem.berkeley.edu/phase_rule.html)

Measurement of the internal energy of a system requires that we have already defined a reference or standard state. This is the state where a component is most stable at normal temperature and pressure. When ΔG is measured at the standard conditions, it is denoted as ΔG°.

For chemists, the standard conditions of a system are a pressure of 1 atm, a temperature equal to 25°C or 298 K, a chemical component concentration of 1 M and consequently a pH $= 0$ (since pH $= -\log[H^+]$ and $[H^+] = 1$ M).

However, standard conditions in biochemistry and in biology are different, since chemical reactions in cells occur at around pH 7. Therefore, standard conditions in biochemistry differ from those in chemistry, which implies that the standard Gibbs free energy within a biological system is denoted as $\Delta G^{\circ\prime}$. Standard conditions in biochemistry and biology are: a pH equal to 7 and a constant water concentration that does not appear in the mathematical definition of the equilibrium constant.

4.6.5 Relation between $\Delta G^{\circ\prime}$ and $\Delta G^{\prime}$

Let us consider the following reaction:

$$aA + bB \underset{2}{\overset{1}{\rightleftarrows}} cC + dD \tag{4.24}$$

The $\Delta G^{\circ\prime}$ of the reaction is

$$\sum \Delta G^{\circ\prime}(\text{products}) - \sum \Delta G^{\circ\prime}(\text{reactants}) \tag{4.25}$$

$$G' = H - TS = U + PV - TS \tag{4.26}$$

where H, S, U, P, and V are the enthalpy, entropy, internal energy, pressure, and volume of the system, respectively.

The first derivative of Equation (4.25) is

$$dG' = dU + P \, dV + V \, dP - T \, dS - S \, dT \tag{4.27}$$

dU can be replaced by $dQ - P \, dV$ where dQ is the variation of heat within the system. Then,

$$dG' = dQ - P \, dV + V \, dP - T \, dS - S \, dT = dQ + V \, dP - T \, dS - S \, dT \tag{4.28}$$

$$dQ = T \, dS \tag{4.29}$$

Replacing Equation (4.29) in Equation (4.28) yields

$$dG' = T \, dS + V \, dP - T \, dS - S \, dT = V \, dP - S \, dT \tag{4.30}$$

At a constant temperature,

$$dG' = V \, dP \tag{4.31}$$

For 1 mol of component,

$$P\overline{V} = RT \tag{4.32}$$

or

$$\overline{V} = RT/P \tag{4.33}$$

Combining Equations (4.32) and (4.33) yields

$$d\overline{G'} = \frac{RT}{P} \, dP \tag{4.34}$$

The integral of Equation (4.34) between $\overline{G}^{\circ'}$ and $\overline{G}'$ and $\overline{P}_0(1 \text{ atm})$ and P yields

$$\int_{\overline{G}^{\circ'}}^{\overline{G}'} d \, \overline{G}' = \int_{\overline{P}_0}^{P} \frac{RT}{P} \, dP \tag{4.35}$$

$$G' - \overline{G}^{\circ'} = RT \log \frac{P}{P_0} = RT \log P \tag{4.36}$$

If we have n mol instead of 1, we have

$$n\overline{G}' = n\overline{G}^{\circ'} + nRT \log P \tag{4.37}$$

4.6.6　Relation between $\Delta G^{\circ'}$ and K_{eq}

Let us consider the following reaction:

$$\alpha A(P_A) + \beta B(P_B) \; \underset{\longleftarrow}{\overrightarrow{}} \; cC(P_C) + dD(P_D) \tag{4.38}$$

$$\sum \Delta G^{\circ'}_{\text{(products)}} - \sum \Delta G^{\circ'}_{\text{(reactants)}} \tag{4.39}$$

$$\Delta G' = c\overline{G}'(C) + d\overline{G}'(D) - \alpha \overline{G}'(A) - \beta \overline{G}'(B) \tag{4.40}$$

$$\Delta G = [c\overline{G}^{\circ\prime}(C) + d\overline{G}^{\circ\prime}(D) - \alpha\overline{G}^{\circ\prime}(A) - \beta\overline{G}^{\circ\prime}(B)]$$
$$+ cRT \log P_C + dRT \log P_D - \alpha RT \log P_A - \beta RT \log P_B \tag{4.41}$$

$$\Delta G' = \Delta G^{\circ\prime} + RT \log \frac{(P_C)^c (P_D)^d}{(P_A)^a (P_B)^b} \tag{4.42}$$

At equilibrium, $\Delta G' = 0$

$$\Rightarrow \Delta G^{\circ\prime} = -RT \log \left(\frac{(P_C)^c (P_D)^d}{(P_A)^a (P_B)^b} \right)_{eq} \tag{4.43}$$

The pressure ratio gives the equilibrium constant K_{eq}, and so we obtain:

$$\Delta G^{\circ\prime} = -RT \log K_{eq} \tag{4.44}$$

R is the ideal gas constant $= 2$ cal mol^{-1} K^{-1}, and T is the temperature in kelvins. Equation (4.44) indicates that $\Delta G^{\circ\prime}$ decreases linearly with ln K_{eq}, with a slope equal to -RT (Figure 4.13). At equilibrium, $\Delta G^{\circ\prime} = 0$, and from the graph one can see that the corresponding value of ln K_{eq} is 4.831×10^{-6}, which corresponds to a K_{eq} equal to 1.

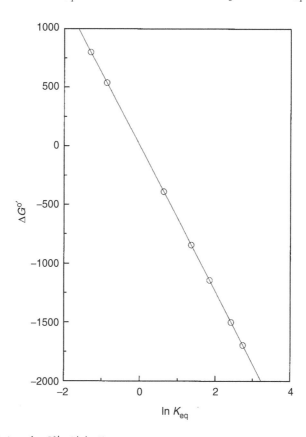

Figure 4.13 Variation of $\Delta G^{\circ\prime}$ with ln K_{eq}.

Equation (4.44) can be written as

$$K_{eq} = e^{-\Delta G^\circ/RT} = 10^{-\Delta G^\circ/2,3RT} \tag{4.45}$$

If $\Delta G^{\circ\prime} < 0$, the exponential will be positive, and K_{eq} higher than 1. Thus, the lower the $\Delta G^{\circ\prime}$, the higher K_{eq} will be.

Therefore, for the reactions having the highest negative values of $\Delta G^{\circ\prime}$ most of the components will be in the products forms.

If $\Delta G^{\circ\prime} > 0$, K_{eq} will be lower than 1. In this case, although we can find some products in the equilibrium state, most of the components will be in the reactant forms. If $\Delta G^{\circ\prime} = 0$, K_{eq} will be equal to 1. Therefore, in the presence of high guanidine concentrations, that of β-galactosidase should increase, and in this case $\Delta G^{\circ\prime}$ would decrease (Figure 4.14).

The $\Delta G^{\circ\prime}$ variation with guanidine chloride concentration can be described by the following equation:

$$\Delta G_U^{\circ\prime} = \Delta G_{H_2O}^{\circ\prime} - mC \tag{4.46}$$

where $\Delta G_U^{\circ\prime}$ is the free energy of unfolding; $\Delta G_{H_2O}^{\circ\prime}$ is the free energy of unfolding in the absence of denaturant, C is the guanidine concentration, and m is a parameter that measures

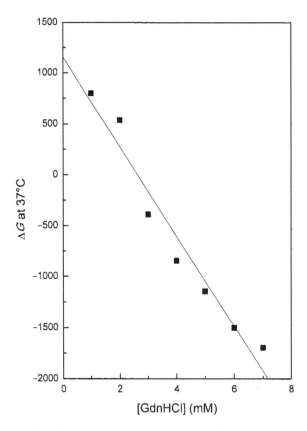

Figure 4.14 Variation of $\Delta G^{\circ\prime}$ of the enzymatic reaction with β-galactosidase as a function of guanidine chloride concentration.

the rate of change in free energy of denaturation with respect to denaturant concentration. m is also called the denaturant susceptibility parameter.

At the midpoint of the denaturation curve, the guanidine concentration is called C_m, and the free energy $\Delta G_U^{0'}$ is zero. This yields

$$\Delta G_{H_2O}^{o'} = mC_m \tag{4.47}$$

Combining Equations (4.46) and (4.47) yields

$$\Delta G_U^{0'} = mC_m - mC \tag{4.48}$$

Thus, plotting $\Delta G_U^{0'}$ as function of C gives a line with a slope equal to m and a y-intercept equal to mC_m. Data of Figure 4.14 yield values of m and C_m equal to 460 cal mol^{-1} mol^{-1} and 2.5 mM, respectively.

Performing the experiment described in Figure 4.10 at 25°C instead of 37°C would decrease the number of collisions between the enzyme and the heteroside, thereby inducing a decrease in the hydrolysis rate. In this case, the general shape of the plot OD $= f($[guanidine chloride]$)$ could be different from that obtained when the experiment was performed at 37°C. In fact, the plot (Figure 4.15) is sigmoidal and clearly shows the presence of a transition

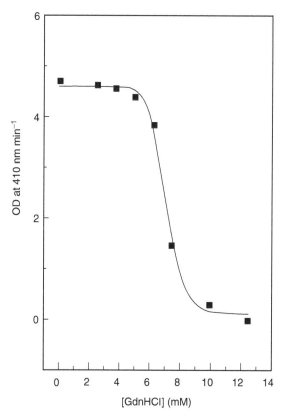

Figure 4.15 Activity variation of β-galactosidase (18.5 μg ml^{-1}) as a function of guanidine concentration. The experiment was performed at 25°C.

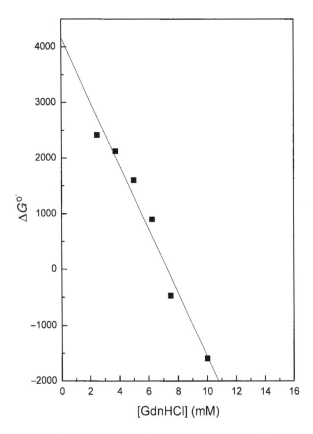

Figure 4.16   Variation of $\Delta G^{o\prime}$ with guanidine chloride concentration at 25°C.

point at 7 mM guanidine. Thus, lowering the temperature shows that there is at least one intermediate structure between folded and unfolded structures of β-galactosidase. Plotting $\Delta G^{o\prime}$ as a function of guanidine chloride concentration allows the values of m and C_m to be determined, i.e., 535 cal mol^{-1} mol^{-1} and 7.8 mM, respectively (Figure 4.16). Results obtained with guanidine chloride show that β-galactosidase activity is dependent on the concentration of the denaturing solvent and that the denaturation rate is temperature-dependent. Also, these experiments show that β-galactosidase unfolding would induce the presence of intermediate states.

4.6.7   Effect of guanidine chloride on hydrolysis kinetics of p-nitrophenyl-β-D-galactoside

Mix in a test tube 0.2 ml of 1 M guanidine with 1.8 ml of enzyme solution stock. The new stock solution is 0.1 M guanidine.

Prepare 11 test tubes with 0.5 ml of 10 mM heteroside and 1.5 ml of pH 7 buffer mixed together, and keep the test tubes at 37°C for 5 min. Then, to each tube add 0.15 ml of the mixture (enzyme–guanidine), mix and let the reaction continue at 37°C. After 0, 1, 2, 3, 4,

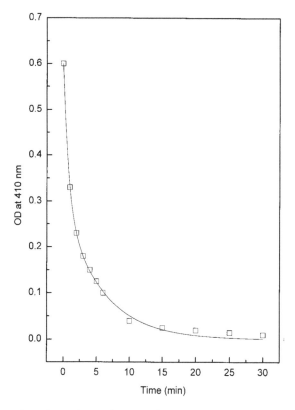

Figure 4.17 Activity of β-galactosidase with time in the presence of 7 mM guanidine chloride. Analysis of the exponential decrease as the sum of two exponentials yields decay times of 2.9 and 14.5 min.

5, 10, 15, 20, 25, and 30 min, add 3 ml of Na_2CO_3, mix, and measure the OD at 405 nm. Finally, plot the OD as a function of time and calculate the decay times of the curve.

What can you conclude about the denaturation process of β-galactosidase?

4.6.7.1 Results

Figure 4.17 displays β-galactosidase activity in the presence of 7 mM of guanidine chloride as a function of time. Analysis of the decay curve yields two decay times equal to 2.9 and 14.5 min. These times reveal mainly that the denaturation process of a protein does not occur in one simple step. The structure and dynamics of the protein should play an important role in the accessibility of the guanidine chloride to the amino acids and so to the unfolding process.

References

Bradford, M.M. (1976). A rapid and sensitive method for the quantitation of microgram quantitites of protein utilizing the principle of protein-dye binding. *Analytical Biochemistry*, **72**, 248–254.

Carpita, N.C. and Gibeaut, D.M. (1993). Structural models of primary cell walls in flowering plants: consistency of molecular structure with the physical properties of the walls during growth. *Plant Journal*, **3**, 1–30.

Murata, T., Hattori, T., Amarume, S., Koichi, A. and Usui, T. (2003). Kinetic studies on endo-beta-galactosidase by a novel colorimetric assay and synthesis of *N*-acetyllactosamine-repeating oligosaccharide beta-glycosides using its transglycosylation activity. *European Journal of Biochemistry*, **270**, 3709–3719.

Stolle-Smits, T., Gerard Beekhuizen, J., Kok, M.T.C., Pijnenburg, M., Jan Derksen, K.R. and Voragen, A.G.J. (1999). Changes in cell wall polysaccharides of green bean pods during development. *Plant Physiology*, **121**, 363–372.

Talbott, L.D. and Ray, P.M. (1992). Molecular size and separability features of pea cell wall polysaccharides: implications for models of primary structure. *Plant Physiology*, **98**, 357–368.

http://www.mpbio.com/product_info.php?cPath=491_1_12&products_id=150039&depth=nested&keywords=beta%20galactosidase

Chapter 5
Starch Hydrolysis by Amylase

Experiment conducted by Quanzeng Wang under the supervision of Professor Nam Sun Wang, Department of Chemical and Biomolecular Engineering, University of Maryland, College Park, MD 20742, USA.

5.1 Objectives

The objective is to study the various parameters that affect the kinetics of α-amylase-catalyzed hydrolysis of starch.

5.2 Introduction

Starchy substances constitute the major part of the human diet for most people in the world, as well as many other animals. They are synthesized naturally in a variety of plants. Examples of plants with a high starch content include corn, potato, rice, sorghum, wheat, and cassava. It is no surprise that all of these are part of what we consume to derive carbohydrates. Similar to cellulose, starch molecules are glucose polymers linked together by α-1,4 and α-1,6 glucosidic bonds, as opposed to the β-1,4 glucosidic bonds for cellulose. In order to make use of the carbon and energy stored in starch, the human digestive system, with the help of the enzyme amylases, must first break down the polymer to smaller assimilable sugars, which are eventually converted to the individual basic glucose units.

Because of the existence of two types of linkages, α-1,4 and α-1,6, different structures are possible for starch molecules. An unbranched, single-chain polymer of 500–2000 glucose subunits with only α-1,4 glucosidic bonds is called *amylose*. On the other hand, the presence of α-1,6 glucosidic linkages results in a branched glucose polymer called *amylopectin*. The degree of branching in amylopectin is approximately one per 25 glucose units in the unbranched segments. Another closely related compound functioning as the glucose storage in animal cells is called *glycogen*, which has one branching per 12 glucose units. The degree of branching and the side-chain length vary from source to source, but in general the more the chains are branched, the more the starch is soluble.

Starch is generally insoluble in water at room temperature. Because of this, starch in nature is stored in cells as small granules, which can be seen under a microscope. Starch granules are quite resistant to penetration by both water and hydrolytic enzymes due to the formation of hydrogen bonds within the same molecule and with other neighboring molecules. However, these inter- and intra-hydrogen bonds can become weak as the temperature of the suspension is raised. When an aqueous suspension of starch is heated, the hydrogen bonds weaken, water is absorbed, and the starch granules swell. This process is commonly called *gelatinization* because the solution formed has a gelatinous, highly viscous consistency. The same process has long been employed to thicken broth in food preparation.

Depending on the relative location of the bond under attack as counted from the end of the chain, the products of this digestive process are dextrin, maltotriose, maltose, glucose, etc. Dextrins are shorter, broken starch segments that form as the result of the random hydrolysis of internal glucosidic bonds. A molecule of maltotriose is formed if the third bond from the end of a starch molecule is cleaved; a molecule of maltose is formed if the point of attack is the second bond; a molecule of glucose results if the bond being cleaved is the terminal one; and so on. The initial step in random depolymerization is the splitting of large chains into various smaller segments. The breakdown of large particles drastically reduces the viscosity of gelatinized starch solution, resulting in a process called *liquefaction* because of the thinning of the solution. The final stages of depolymerization are mainly the formation of mono-, di-, and tri-saccharides. This process is called *saccharification*, due to the formation of saccharides.

Since a wide variety of organisms, including humans, can digest starch, α-amylase is obviously widely synthesized in nature, as opposed to cellulase. For example, human saliva and pancreatic secretion contain a large amount of α-amylase for starch digestion. The specificity of the bond attacked by α-amylases depends on the sources of the enzymes. Currently, two major classes of α-amylases are commercially produced through microbial fermentation. Based on the points of attack in the glucose polymer chain, they can be classified into two categories, liquefying and saccharifying. Because the bacterial α-amylase to be used in this experiment randomly attacks only the α-1,4 bonds, it belongs to the liquefying category. The hydrolysis reaction catalysed by this class of enzymes is usually carried out only to the extent that, for example, the starch is rendered soluble enough to allow easy removal from starch-sized fabrics in the textile industry. The paper industry also uses liquefying amylases on the starch used in paper coating where breakage into the smallest glucose subunits is actually undesirable (one cannot bind cellulose fibers together with sugar!)

On the other hand, the fungal α-amylase belongs to the saccharifying category and attacks the second linkage from the nonreducing terminals (i.e., the C4 end) of the straight segment, resulting in the splitting off of two glucose units at a time. Of course, the product is a disaccharide called maltose. The bond breakage is thus more extensive in saccharifying enzymes than in liquefying enzymes. The starch chains are literally chopped into small bits and pieces. Finally, the amyloglucosidase (also called glucoamylase) component of an amylase preparation selectively attacks the last bond on the nonreducing terminals. The type to be used in this experiment can act on both the α-1,4 and the α-1,6 glucosidic linkages at a relative rate of 1:20, resulting in the splitting off of simple glucose units into the solution. Fungal amylase and amyloglucosidase may be used together to convert starch to simple sugars. The practical applications of this type of enzyme mixture include the production of corn syrup and the conversion of cereal mashes to sugars in brewing.

Thus, it is important to specify the source of enzymes when the actions and kinetics of the enzymes are compared. In the present work, α-amylases from microbial origin will be employed. The effects of temperature, pH, substrate concentration, and inhibitor concentration on the kinetics of amylase-catalysed reactions will be studied. Finally, it is important and interesting to know that the more accurate name for α-amylase is 1,4-α-D-glucan-glucanohydrolase (EC 3.2.1.1); for β-amylase, 1,4-α-D-glucanmaltohydrolase (EC 3.2.1.2); and for amyloglucosidase, exo-1,4-α-glucosidase or 1,4-α-D-glucan glucohydrolase (EC 3.2.1.3).

5.3   Materials

- α-Amylase (EC 3.2.1.1): Type VIII-A, from barley malt [9000-90-2], Sigma A-2771, lot 124H0151. Prepare a 20 g l^{-1} solution
- Sigma's definition: 1 unit liberates 1.0 mg of maltose from starch in 3 min at pH 6.9 at 20°C
- Purity: 1.1 units α-amylase per milligram of solid, 1.7 units β-amylase per milligram of solid at pH 4.8
- Starch: potato, soluble [9005-84-9], Sigma S-2630, lot 44H0156. Prepare a 20 g l^{-1} solution
- Maltose: maltose, hydrate, Sigma M-5885, lot 37F0869
- Glucose: D-(+)-glucose, Sigma G-7021, lot 76H0043
- HCl stopping solution, 0.1N HCl
- Iodine reagent stock solution (in aqueous solution). See Note 1
- Iodine: 5 g l^{-1}
- KI: 50 g l^{-1}
- KH_2PO_4 (monobasic phosphate) (FW = 136.1)
- $K_2HPO_4 \cdot 3H_2O$ (dibasic phosphate) (FW = 228.23)
- $CaCl_2 \cdot 2H_2O$, 0.1 M solution
- Reagents for the analysis of reducing sugars
- Room temperature = 22.5°C

5.4   Procedures and Experiments

5.4.1   *Preparation of a 20 g l^{-1} starch solution*

Mix 20 g of soluble potato starch in approx. 50 ml of cold water. While stirring, add the slurry to approx. 900 ml of gently boiling water in a large beaker. Mix well and cool the gelatinized starch solution to room temperature. Add more water to bring the total volume up to 1 l. Put a few drops of the starch solution on a glass plate. Add 1 drop of the iodine reagent; you should see that a deep blue color is developed. The blue color indicates the presence of starch in the solution.

5.4.2   *Calibration curve for starch concentration*

Add 0.1 ml of sample to 1 ml of 0.1 N HCl, and then add 0.2 ml of the mixture to 2 ml of iodine solution. Finally, measure the absorbance with the spectrophotometer at 620 nm (note 2). The resulting data and plot of starch concentration as a function of absorbance are shown in Table 5.1 and Figure 5.1, respectively.

5.4.2.1   *Conclusion*

The linear plot (Figure 5.1) indicates that the intensity of the blue color is directly related to starch concentration.

Table 5.1 Absorbance at 620 nm for different starch concentrations

Starch concentration (g l^{-1})	Volume of 10 g l^{-1} Starch (ml)	Volume of water (ml)	Absorbance at 620 nm
0	0	2	0.003
1	0.2	1.8	0.080
2	0.4	1.6	0.167
3	0.6	1.4	0.236
4	0.8	1.2	0.322
5	1	1	0.395
6	1.2	0.8	0.470
7	1.4	0.6	0.553
8	1.6	0.4	0.618
9	1.8	0.2	0.697
10	2	0	0.784

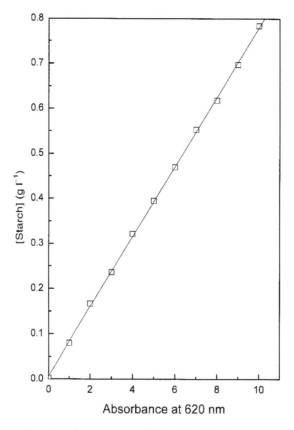

Figure 5.1 Calibration plot of starch concentration followed by absorbance at 620 nm.

5.4.3 Calibration curve for sugar concentration

Add 1 ml of DNS reagent to 1 ml of maltose (or glucose) sample. Heat the mixture to 100°C for 10 min. Then, add 0.33 ml of a 40% potassium sodium tartrate. Cool to room temperature in a cold water bath, and measure the absorbance at 575 nm.

The results are displayed in Table 5.2 and Figure 5.2.

Table 5.2 Absorbance variation at 575 nm as a function of glucose and maltose

Maltose (or glucose) concentration (g l^{-1})	Volume of 10 g l^{-1} maltose (or glucose) (ml)	Volume of water (ml)	Absorbance of maltose at 575 nm	Absorbance of glucose at 575 nm
0	0	1	0.015	0.020
1	0.1	0.9	0.540	0.576
2	0.2	0.8	1.068	1.166
3	0.3	0.7	1.534	1.708
4	0.4	0.6	1.993	2.196
5	0.5	0.5	2.322	2.461

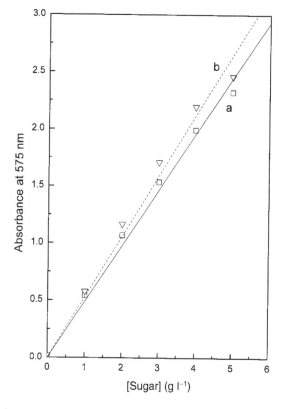

Figure 5.2 Optical-density variation at 575 nm as a function of maltose (a) and glucose (b) concentration.

5.4.3.1 Conclusion

Both maltose and glucose can be reliably measured, albeit the color intensities are slightly different.

5.4.4 Effect of pH

The experiment is carried out at room temperature (22.5°C).

1   Prepare 0.25 M pH buffer solutions ranging from pH 0.5 to 9. (Note that phosphate buffer is only good for pH = 4.5–9 due to the dissociation constant.) Before coming to the lab, review how to make a pH buffer solution in a freshman chemistry textbook and calculate the relative amounts of KH_2PO_4 (monobasic phosphate) and $K_2HPO_4 \cdot 3H_2O$ (dibasic phosphate) needed to make these phosphate buffer solutions.
2   Add an equal volume of one of the above buffer solutions to 2.0 ml of the 20 g l^{-1} starch solution prepared in step 1. The resulting solution should contain 10 g l^{-1} of starch in a buffered environment.
3   Add 0.2 ml of the barley amylase to 2 ml of the 10 g l^{-1} buffered starch solution, and keep the reaction going for 3 min.
4   After 3 min, stop the enzymatic reaction by adding 0.1 ml of the reacted starch solution to 1 ml of the HCl stopping solution (0.1 N).
5   Then, add 0.2 ml of the above mixture to 2 ml of iodine solution to develop the color. Shake and mix. The solution should turn deep blue if there is any residual, unconverted starch present in the solution. The solution is brown-red for partially degraded starch but clear for totally degraded starch.
6   Measure the absorbance at 620 nm. For an absorbance value, the equivalent starch concentration is calculated from Figure 5.1.
7   Carry out the same procedure for the other starch solutions buffered at different pHs. Use your time wisely; all the solutions can be handled simultaneously if you are familiar with the procedure. Slightly stagger the sequential sample withdrawal so that there is enough time for sample preparation and handling.

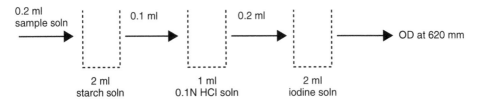

Experimental data are shown in Table 5.3, and the results are displayed in Table 5.3 and Figure 5.3. The actual pH in Table 5.3 corresponds to the pH of the measurement.

5.4.4.1 Notes

Water is added instead of amylase in control experiments. Our enzymatic activity definition is: 1 unit of amylase hydrolyses 1.0 mg of starch in 3 min at pH 6.6 at 20°C.

Table 5.3 Data for measuring the barley amylase activity on starch at different pH

Actual pH	Volume of HCl (μl)	Volume of water (μl)	Volume of 20 g l^{-1} starch (ml)	Absorbance at 620 nm	Corresponding starch concentration (g l^{-1})	Starch mass (mg)	Specific activity (units mg^{-1})
				With amylase			
0.78	2000 (1 M)	0	2	0.570	7.3	0.7	10
0.97	900 (1 M)	1100	2	0.542	6.9	0.7	1.2
1.50	200 (1 M)	1800	2	0.576	7.4	0.7	1.0
2.90	21 (0.1 M)	1979	2	0.419	5.3	0.5	2.1
3.38	7 (0.1 M)	1993	2	0.378	4.8	0.5	2.4

Actual pH	pH of 0.25 M buffer	Volume of 0.25 M buffer (ml)	Volume of water (ml)	Volume of 20 g l^{-1} starch (ml)	Absorbance at 620 nm	Starch concentration (g l^{-1})	Starch mass (mg)	Specific activity (units mg^{-1})
4.6	4.45	0.4	0.6	1	0.195	2.4	0.2	3.7
5.6	5.48	0.4	0.6	1	0.376	4.8	0.5	2.4
6.6	6.52	0.4	0.6	1	0.528	6.7	0.7	1.3
7.6	7.58	0.4	0.6	1	0.607	7.8	0.8	0.7
8.4	8.5	0.4	0.6	1	0.66	8.4	0.8	0.4
9.1	9.3	0.4	0.6	1	0.638	8.2	0.8	0.5

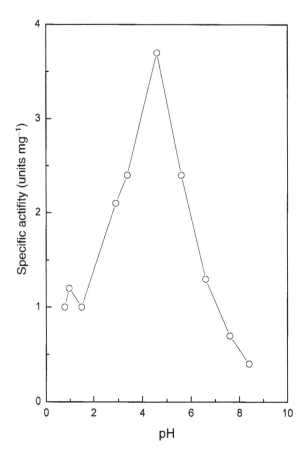

Figure 5.3 Amylase activity with pH.

The pHs of the blue solutions are adjusted with 0.1 and 1 M HCl. In the experimental conditions described here, in 0.1 ml sample, the initial starch at the beginning of the experiment is 0.91 mg, and the initial amylase is 0.18 mg.

5.4.4.2 Conclusion

For the enzyme employed (α-amylase from barley), the optimum pH is about 4.5. As the pH moves away from the optimum value in either direction, the enzyme activity decreases. Note that from our experience (data not shown), the optimum pH depends on the source of the enzyme.

5.4.5 Temperature effect

Note: the solution can be prepared together and then separated to aliquots of 2 ml.

1 Obtain hot water from either a faucet or a hot temperature bath. Adjust the temperatures of the temporary water baths in 500 ml beakers so that they range from 30 to 90°C in increments of 10°C.

2   Prepare the starch substrate by diluting the 20 g l^{-1} starch solution prepared in step 1 with an equal volume of pH 7.0 phosphate buffer solution. This results in a working starch concentration of 10 g l^{-1}. Add 2 ml of the starch solution to each of the test tubes.
3   Allow the temperature of each of the starch solutions to reach equilibrium with that of the water bath.
4   Add 0.1 ml of the barley amylase to 2 ml of the thermostatted 10 g l^{-1} buffered starch solution.
5   Let the enzyme act for 3 min.
6   Add 0.1 ml of the reacted starch solution to 1 ml of the HCl stopping solution (0.1 N)
7   Add 0.2 ml of the above mixture to 2 ml of iodine solution to develop the color.
8   Measure the absorbance at 620 nm.

Follow exactly the procedure outlined in step 4 (experiment at different pH). In 0.1 ml sample, we have an initial starch value of 0.95 mg and initial amylase of 0.10 mg.

Tables 5.4 and 5.5 Show, respectively, the experimental procedure and the results obtained in the presence of amylase and of water (control experiment), and Figures 5.4 and 5.5 show

Table 5.4   Experimental procedure and results obtained for starch hydrolysis in the presence of amylase at different temperatures and at pH 7

					With amylase			
Temperature (°C)	pH of 0.25 M buffer	Volume of 0.25 M buffer (ml)	Volume of water (ml)	Volume of 20 g l^{-1} starch (ml)	Absorbance at 620 nm	Starch concentration (g l^{-1})	Starch mass (mg)	Specific activity (units mg^{-1})
30	6.52	0.4	0.6	1	0.551	7.0	0.7	2.5
40	6.52	0.4	0.6	1	0.429	5.5	0.5	4.0
50	6.52	0.4	0.6	1	0.078	0.9	0.1	8.6
60	6.52	0.4	0.6	1	0.040	0.4	0.0	9.1
70	6.52	0.4	0.6	1	0.464	5.6	0.6	3.9
80	6.52	0.4	0.6	1	0.666	8.5	0.9	1.0
90	6.52	0.4	0.6	1	0.673	8.6	0.9	0.9

Table 5.5   Experimental procedure and results obtained for starch hydrolysis in the presence of water at different temperatures

					With water		
Temperature (°C)	pH of 0.25 M buffer	Volume of 0.25 M buffer (ml)	Volume of water (ml)	Volume of 20 g l^{-1} starch (ml)	Absorbance at 620 nm (water control)	Starch concentration (g l^{-1})	Starch mass (mg)
30	6.52	0.4	0.6	1	0.629	8.0	0.8
40	6.52	0.4	0.6	1	0.638	8.2	0.8
50	6.52	0.4	0.6	1	0.662	8.5	0.8
60	6.52	0.4	0.6	1	0.646	8.3	0.8
70	6.52	0.4	0.6	1	0.625	8.0	0.8
80	6.52	0.4	0.6	1	0.697	8.9	0.9
90	6.52	0.4	0.6	1	0.664	8.5	0.8

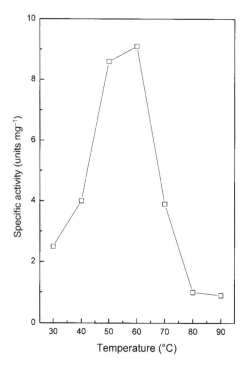

Figure 5.4   Determination of the optimum temperature of amylase activity.

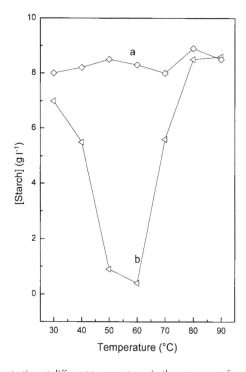

Figure 5.5   Starch concentration at different temperatures in the presence of amylase (b) and of water (a).

the specific activity and the corresponding starch concentration, respectively, at different temperatures.

5.4.5.1 Conclusion

For the enzyme employed, the highest conversion occurs at about 50°C. As temperature increases, the intrinsic enzyme activity increases; however, enzyme deactivation also becomes significant. Note that, as in the optimum pH, the optimum temperature depends on the source of the enzyme.

5.4.6 Effect of heat treatment at 90° C

Note: the solution can be prepared together and then separated to aliquots of 5 ml.

1   Put 0.4 ml of the amylase solution in each of 11 tubes, and then place the tubes into a 90°C water bath.
2   Take out the first test tube after a certain time (min) and quickly bring it to room temperature in a cold water bath. Remove the second test tube after x min, the third after y min, and so on. Add 2 ml of 10 g l^{-1} buffered/nonbuffered starch at room temperature to each tube, and continue the enzymatic reaction for 3 min.
3   Add 0.1 ml of the reacted starch solution to 1 ml of the HCl stopping solution (0.1 N)
4   Add 0.2 ml of the above mixture to 2 ml of iodine solution to develop the color, then measure the absorbance at 620 nm.

Do not forget to follow exactly the procedure outlined in the experiment at different pH. In 0.1 ml sample, initial starch: 0.83 mg and initial amylase: 0.33 mg. $[Ca^{2+}] = 0.08$ M. The pH of 0.25 M buffer used is 6.52, which is the pH (6.5) of the experiment.

The experiment should be performed under three conditions: with buffer (Table 5.6), without buffer (only water is added) (Table 5.7), and in CaCl$_2$ solution to investigate the heat stabilization of the enzymes in the presence of Ca^{2+} ions (Table 5.8).

Figures 5.6 and 5.7 show, respectively, amylase activity and starch concentration vs. heating time at 90°C in the presence and absence of buffer.

Figures 5.8 and 5.9 show, respectively, amylase activity and starch concentration vs. heating time at 90°C in the presence and absence of CaCl$_2$.

Table 5.6 Specific activity data for amylase activity after heat treatment of buffered enzyme at 90°C

Heat time (min)	Volume of 0.25 M buffer (ml)	Volume of water (ml)	Volume of 20 g l^{-1} starch (ml)	Absorbance At 620 nm	Starch concentration (g l^{-1})	Starch mass (mg)	Specific activity (units mg^{-1})
0.0	0.4	0.6	1	0.399	5.1	0.5	1.0
0.3	0.4	0.6	1	0.492	6.3	0.6	0.6
0.7	0.4	0.6	1	0.511	6.5	0.7	0.5
1.0	0.4	0.6	1	0.549	7.0	0.7	0.4
2.0	0.4	0.6	1	0.613	7.8	0.8	0.1
3.5	0.4	0.6	1	0.619	7.9	0.8	0.1
5.0	0.4	0.6	1	0.621	7.9	0.8	0.1

Table 5.7 Specific activity data of amylase activity after heat treatment of buffered enzyme at 90°C

Heat time (min)	Volume of 0.25 M buffer (ml)	Volume of water (ml)	Volume of 20 g l^{-1} starch (ml)	Absorbance at 620 nm	Starch concentration (g l^{-1})	Starch mass (mg)	Specific activity (units mg^{-1})
0.0	0	1	1	0.401	5.1	0.5	1.0
0.3	0	1	1	0.489	6.2	0.6	0.6
0.7	0	1	1	0.517	6.6	0.7	0.5
1.0	0	1	1	0.58	7.4	0.7	0.3
2.0	0	1	1	0.587	7.5	0.8	0.2
3.5	0	1	1	0.592	7.6	0.8	0.2
5.0	0	1	1	0.598	7.6	0.8	0.2

Table 5.8 Specific activity data of amylase after heat treatment of enzyme at 90°C in the absence of buffer and presence of CaCl$_2$

Heat time (min)	Volume of 0.2 M CaCl$_2$	Volume of water (ml)	Volume of 20 g l^{-1} starch (ml)	Absorbance at 620 nm	Starch concentration (g l^{-1})	Starch mass (mg)	Specific activity (units mg^{-1})
0.0	0.2	0.8	1	0.41	5.2	0.5	0.9
0.3	0.2	0.8	1	0.495	6.3	0.6	0.6
0.7	0.2	0.8	1	0.545	7.0	0.7	0.4
1.0	0.2	0.8	1	0.606	7.8	0.8	0.2
2.0	0.2	0.8	1	0.629	8.0	0.8	0.1
3.5	0.2	0.8	1	0.648	8.3	0.8	0.0
5.0	0.2	0.8	1	0.651	8.3		0.0

5.4.6.1 Conclusion

The rate of deactivation is temperature-dependent. At 90°C (which data are shown), this enzyme becomes completely deactivated in ~4 min.

5.4.7 Kinetics of starch hydrolysis

This experiment is performed under two conditions: in the absence and presence of 0.25 M buffer (pH 6.52).

In the absence of buffer, 1.2 ml of water is added to 1.2 ml of 20 g l^{-1} starch. Then, 0.1 ml of the barely enzyme is added to the 2.4 ml, 10 g l^{-1} starch solution.

In the presence of buffer, 2.4 ml of 0.25 M buffer is mixed with 3.6 ml of water and 6 ml of 20 g l^{-1} starch. Then, 0.5 ml of the barely enzyme is added to the 5 ml, 10 g l^{-1} starch solution. Then, the procedure follows that described in the heat-treatment section. In the reaction mixture, the initial starch concentration is 9.6 mg ml^{-1}, and the initial amylase concentration is 0.80 mg ml^{-1}.

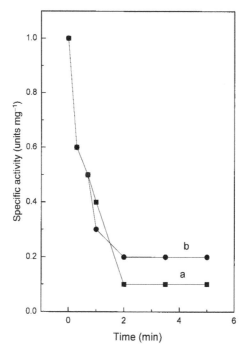

Figure 5.6 Specific activity of amylase as a function of heating time at 90°C, in the presence (a) and absence of buffer (b).

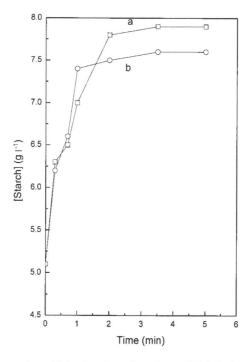

Figure 5.7 Starch concentration with heating time of amylase at 90°C, in the presence (a) and absence (b) of buffer.

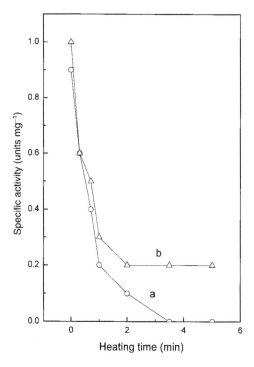

Figure 5.8 Amylase activity vs. heating time at 90°C, in the presence (a) and absence (b) of $CaCl_2$.

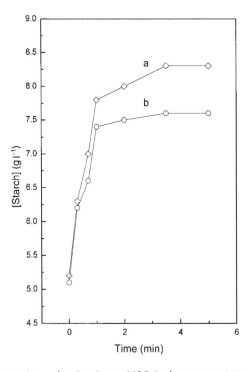

Figure 5.9 Starch concentration vs. heating time at 90°C, in the presence (a) and absence (b) of $CaCl_2$.

Table 5.9 Experimental data and results of starch hydrolysis with time in the absence of buffer (pH 6.2 + 0.1)

Time (min)	Absorbance at 620 nm	Starch concentration (g l^{-1})
0	0.639	8.2
2	0.566	7.2
4	0.507	6.5
6	0.449	5.7
9	0.355	4,5
12	0.238	3.0
17	0.095	1.1
21	0.049	0.6
25	0.032	0.3

Table 5.10 Experimental data and results of starch hydrolysis with time in the presence of buffer (pH 6.4 + 0.1)

Time (min)	Absorbance at 620 nm	Starch concentration (g l^{-1})	Absorbance at 575 nm	Equivalent maltose concentration (g l^{-1})	Equivalent glucose concentration (g l^{-1})
0.1	0.629	8.0	0.008	2.2	1.2
2	0.588	7.5	0.008	2.2	1.2
4	0.508	6.5	0.008	2.2	1.2
6	0.441	5.6	0.008	2.2	1.2
9	0.363	4.6	0.008	2.2	1.2
12	0.256	3.2	0.008	2.2	1.2
17	0.115	1.4	0.01	2.8	1.6
21	0.114	1.4	0.011	3.0	1.8
25	0.112	1.4	0.011	3.0	1.8

The experimental data and results obtained are shown in Tables 5.9 and 5.10, and Figures 5.10 and 5.11.

5.4.8 Effect of inhibitor (CuCl₂) on the amylase activity

The experiment is performed in the absence of buffer, since addition of $CuCl_2$ to the medium in the presence of buffer induces the formation of a precipitate. 0.5 ml of the barely enzyme is mixed to 0.6 ml of 0.2 M $CuCl_2$, 0.6 ml water and 1.2 ml of 20 g l^{-1} starch. Then, the procedure follows that described in the heat treatment section.

Table 5.11 shows the data obtained in the presence of 0.1 M $CuCl_2$, and Figure 5.12 shows the variation in starch concentration in the absence and presence of $CuCl_2$.

5.4.9 Effect of amylase concentration

The experiment is carried out at 25°C and at pH 6.52 (pH of the buffer). Add the barley amylase to the starch solution and let the reaction continue for 3 min. Then, add 0.1 ml of the

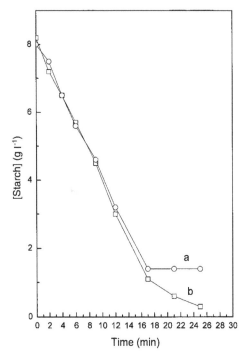

Figure 5.10   Variation in starch concentration with time in the absence (b) and presence (a) of buffer.

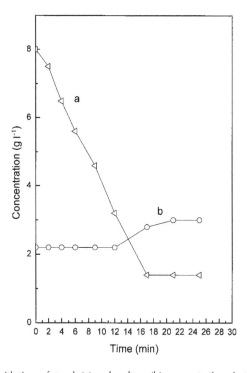

Figure 5.11   Variation with time of starch (a) and maltose (b) concentrations in the presence of buffer.

Table 5.11 Experimental data obtained for starch hydrolysis with amylase at room temperature in the presence of 0.1 M $CuCl_2$

Time (min)	Absorbance at 620 nm	Starch concentration (g l^{-1})
1	0.620	7.9
2	0.602	7.7
4	0.583	7.5
6	0.566	7.2
9	0.535	6.8
12	0.511	6.5
18	0.474	6.0
21	0.461	5.9
25	0.452	5.8

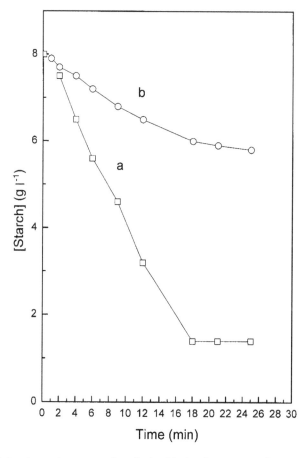

Figure 5.12 Variation in starch concentration obtained in the absence (a) and presence (b) of $CuCl_2$.

reacted starch solution to 1 ml of the HCl stopping solution (0.1 N). Finally, add 0.2 ml of the above mixture to 2 ml iodine solution, mix to develop the color, and measure the absorbance at 620 nm. Do not forget to pay attention to the details given in the heat-treatment section.

Table 5.12 and Figure 5.13 display, respectively, the experimental data and results and starch concentration variation with amylase concentration.

Table 5.12 Experimental data obtained for starch hydrolysis with amylase at 25°C with different amylase concentrations

Enzyme volume (ml)	Enzyme concentration (mg ml^{-1})	Volume of 0.25 M buffer (ml)	Volume of water (ml)	Volume of 20 g l^{-1} starch (ml)	Absorbance at 620 nm	Starch concentration (g l^{-1})	Starch mass (mg)	Specific activity (units mg^{-1})
0	0.0	0.4	0.6	1	0.622	8.0	0.8	
0.1	1.0	0.4	0.6	1	0.518	6.6	0.7	3.1
0.2	1.8	0.4	0.6	1	0.418	5.3	0.5	2.1
0.3	2.6	0.4	0.6	1	0.353	4.5	0.4	1.6
0.4	3.3	0.4	0.6	1	0.279	3.5	0.4	1.4
0.5	4.0	0.4	0.6	1	0.222	2.8	0.3	1.3

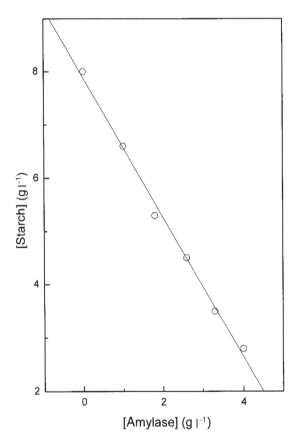

Figure 5.13 Variation in starch concentration with amylase concentration.

5.4.9.1　Conclusion

Starch conversion in a fixed time relates linearly to the enzyme concentration.

5.4.10　Complement experiments that can be performed

5.4.10.1　Experiments with enzymes from different sources

The experiments described above can be repeated with enzymes from different sources, such as fungal amylase, amyloglucosidase, and a mixture of equivalent concentrations of three enzymes, bacterial amylase, fungal amylase, and amyloglucosidase.

5.4.10.2　Sequential enzymatic treatment (corn syrup production)

In making industrial sugars, e.g., corn syrup, large gelatinized starch molecules are first chopped into smaller dextrins with the help of bacterial amylase. The liquefaction step is followed by saccharification with either fungal amylase or amyloglucosidase, depending on the end use of the sugar. These sequential enzymatic treatment steps will be simulated in this part of the experiment

Add 0.5 ml of the bacterial amylase solution to 50 ml of the 20 g l^{-1} nonbuffered starch solution prepared in Step 1. Periodically place a few drops of the reaction mixture on a glass plate and add one drop of the iodine reagent. The color should finally turn red, indicating the total conversion of starch to dextrin. This liquefaction step should last for approximately 10 min.

When the process of liquefaction is complete, adjust the pH of the starch solution to 4.7 with 1 N HCl. Filter the starch solution if it is turbid. Separate the solution into two equal parts. To the first starch solution, add 0.5 ml of amyloglucosidase; to the second solution, add 0.5 ml of fungal amylase solution.

Measure the sugar concentrations periodically. Note that you need to use the appropriate calibration curves because one is maltose, and the other is glucose. Also, do not forget to reference your observation to the initial absorbance at the start of the saccharification process so that the increase in the sugar concentration can be correctly measured. This saccharification step should last for about 30–60 min.

Taste the two sugar solutions and compare the sweetness. See Note 3.

5.4.10.3　Inhibition

Add hydrogen peroxide at a concentration of 0.5 g l^{-1} to the buffered starch solution at pH 7.0. If time permits, try hydrogen peroxide at a concentration of 1.0 g l^{-1}.

5.4.11　Notes

1　Dilute the stock solution 1:100 to obtain a working solution. Other dilutions may be used, depending on the enzyme activity. Iodine does not dissolve well in water. Iodine (I_2) alone or iodide (I^-) alone does not color starch. It is the tri-iodide complex (I_3^-, formed by $I_2 + I^-$) that gives the blue color when it is incorporated into the coil structure of starch.

2 Remember to take care of the background absorbance caused by the colored iodine solution. The true absorbance should be roughly proportional to the starch concentration. The enzyme solution may have to be diluted first if all the starch present in the sample is digested, and the entire color disappears in 10 min. The most reliable results are obtained when the decrease in the absorbance is approximately 20–70% of the absorbance of the original, undigested starch solution. To measure the amount of starch digested, you need to know the absorbance corresponding to the initial undigested starch solution by following the same procedure with a sample in which plain water in lieu of the enzyme solution is added to the starch solution.

3 Be sure you do not contaminate your sugar solution during the various stages of the reaction. Do not taste the sugar solutions and risk your health if you are not confident about your lab techniques. Nevertheless, since the reagents used are chemical, avoid tasting them. Do not expect others to trust your results if you cannot even convince yourself. All glassware used in this biochemical engineering laboratory should always be much cleaner than the eating utensils on your dinning table. The reagent or analytical grade chemicals we use are also much purer than the food grade ones. Furthermore, as sterility will be stressed in the later part of the course when microorganisms are introduced, the glassware used then should be thoroughly aseptic, certainly cleaner than your finger. If you have hesitation in eating and drinking from any of the glassware or spatula that you use, I suggest that you get into the habit of *really* cleaning them before using them in the experiment. Although I am not encouraging you to go around and lick everything in sight, you should develop a good, aseptic laboratory habits so that you know you will not hesitate to do so if needed. The small amount of HCl added to adjust the pH to 4.7 should not affect you at all; many carbonated drinks are much more acidic than this.

4 Do as many experiments as you wish or as time and supplies/materials permit. You can effectively cover all the procedures by teaming up with a few other classmates and exchanging data at the end of the lab period. (Be sure to give the proper credit, or blame for that matter, to your lab partners. Also, be sure you know what your lab partners have done.) However, you must prepare your own lab report.

References

Bailey, J.E. and Ollis, D.F. (1986). *Biochemical Engineering Fundamentals* (2nd edn), Chapter 3, McGraw-Hill, New York.

Sandstedt, R.M., Kneen, E. and Blish, M.J. (1939). A standardized Wohlgemuth procedure for alpha-amylase activity. *Cereal Chemistry* **16**, 712–723.

Chapter 6
Determination of the pK of a Dye

6.1 Definition of pK

Let us consider the following protonation reaction:

$$A + H^+ \Leftrightarrow AH^+ \tag{6.1}$$

The protonation or association constant K_a of the complex is

$$K_a = [AH^+]/[A][H^+] \tag{6.2}$$

$[H^+]$, $[A]$, and $[AH^+]$ characterize the various concentrations:

$$\log K_a = \log([AH^+]/[A]) - \log[H^+] \tag{6.3}$$

$$\log K_a = \log([AH^+]/[A]) + pH \tag{6.4}$$

When A is protonated for 50%, that is, $[AH^+] = [A]$, we reach a pH called pK. Thus:

$$\log K_a = \log 1 + pK \tag{6.5}$$

and

$$pK = \log K_a \tag{6.6}$$

6.2 Spectrophotometric Determination of pK

A pH-sensitive dye induces two different absorption spectra for the protonated (AH^+) and deprotonated (A) forms. This means that at low pHs, we observe an absorption spectrum of the protonated form only, while at a high pHs only the absorption spectrum of the deprotonated form is recorded. However, this is not always true, since many dyes can absorb in both forms, protonated and deprotonated. The absorption spectra maxima of the two forms are different. The protonated form displays an absorption spectrum with a peak that is shifted toward short wavelengths in comparison with the absorption spectrum of the deprotonated form of the dye.

Therefore, one method to determine the pK of a dye is to plot the peak position of the absorption spectrum as a function of pH. A sigmoid curve is obtained; the pK is equal to the pH at the inflection point.

A second method that allows determining the pK of a dye is to plot the optical densities as a function of the pH at two wavelengths λ_1 and λ_2, which are, respectively, the absorption peaks of the protonated and deprotonated forms. Thus, the acidic form of the dye absorbs strongly at wavelength λ_1, and the basic form absorbs strongly at λ_2. At a low pH, we are going to monitor at λ_1, mainly absorbance of the acidic form of the dye. Increasing the pH will increase the concentration of the deprotonated form and decrease that of the protonated form. Therefore, the optical density recorded at λ_1 decreases when the pH increases. At a low pH and at λ_2, we will record mainly the absorbance of the deprotonated form of the dye. However, since the deprotonated form is barely present at low pHs, a very weak optical density is recorded. Increasing the pH induces the formation of deprotonated species, and so we shall observe an increase in the optical density at λ_2. pK is equal to the pH where the concentrations of the protonated and deprotonated forms of the dye are equal. In terms of the experiment, this corresponds to the pH where the absorbance of each form is half its maximum, i.e., at the inflection points of the two curves.

Also, a third method exists. This involves calculating the concentrations of both the protonated and deprotonated forms at different pHs. The pK will correspond to the pH where the concentrations of both forms are equal. The derivation of the mathematical equation used to determine the pK can be obtained as follows. The variation of the optical density of the dye with the pH is sigmoidal. At low pHs in the presence of the protonated form only, the optical density $(OD_{(AH)})$ is

$$OD_{AH} = \varepsilon_{AH} C_{AH} l \tag{6.7}$$

where $\varepsilon_{.(AH)}$, $C_{(AH)}$, and l are the molar extinction coefficient of the protonated form of the dye, its concentration, and the optical path length, respectively.

At high pHs, we observe the optical density of the deprotonated form only $(OD_{(A)})$, which is

$$OD_A = \varepsilon_{.A} C_A l \tag{6.8}$$

where $\varepsilon_{.(A)}$, $C_{(A)}$, and l are the molar extinction coefficient of the deprotonated form of the dye, its concentration, and the optical path length, respectively.

When the dye is completely protonated or deprotonated, each of the concentrations $C_{(AH)}$ and $C_{(A)}$ will be equal to the total concentration, C, of the dye. At an intermediate pH where both protonated and deprotonated forms are present, the optical density is

$$OD_{obs} = OD_{AH} + OD_A = \varepsilon_{.AH} C_{AH} l + \varepsilon_{.A} C_A l \tag{6.9}$$

where the total concentration C is

$$C = C_{AH} + C_A \tag{6.10}$$

Equation (6.9) can also be written as

$$OD_{obs} = \varepsilon_{.AH}(C - C_A)l + \varepsilon_{.A} C_A l \tag{6.11}$$

$$= \varepsilon_{.AH} C l - \varepsilon_{.AH} C_A l + \varepsilon_{.A} C_A l \tag{6.12}$$

$$= OD_{AH} C_{(A} l (\varepsilon_{.A} - \varepsilon_{.AH}) \tag{6.13}$$

Equation (6.13) can also be written as

$$OD_{obs} = OD_{AH} + X_A C l (\varepsilon_{.A} - \varepsilon_{.AH}) \tag{6.14}$$

where $X_{(A)}$ is the fraction of the protonated form of the dye.

$$OD_{obs} = OD_{AH} + X_A Cl\varepsilon_{.A} - X_A Cl\varepsilon_{.AH}$$

$$= OD_{AH} + X_A OD_A - X_A OD_{AH}$$

$$= OD_{AH} + X_A[OD_A - OD_{AH}] \tag{6.15}$$

$$OD_{obs} - OD_{AH} = X_A[OD_A - OD_{AH}]$$

$$X_A = \frac{OD_{obs} - OD_{AH}}{OD_A - OD_{AH}} \tag{6.16}$$

$$X_{AH} = 1 - X_A = \frac{OD_A - OD_{obs}}{OD_A - OD_{AH}} \tag{6.17}$$

One can replace $[AH^+]$ and $[A]$ in Equation (6.4) with the fractions $X_{(AH)}$ and $X_{(A)}$ expressed as optical densities (Equations (6.16) and (6.17)).

$$\frac{[AH^+]}{[A]} = \frac{OD_A - OD_{obs}}{OD_A - OD_{AH}} : \frac{OD_{obs} - OD_{AH}}{OD_A - OD_{AH}} = \frac{OD_A - OD_{obs}}{OD_{obs} - OD_{AH}} \tag{6.18}$$

Combining Equations (6.18) and (6.4) yields

$$\log K_a = \log \frac{OD_A - OD_{obs}}{OD_{obs} - OD_{AH}} + pH \tag{6.19}$$

$$pH = pK - \log \frac{OD_A - OD_{obs}}{OD_{obs} - OD_{AH}} \tag{6.20}$$

$$pH = pK + \log \frac{OD_{obs} - OD_{AH}}{OD_A - OD_{obs}} \tag{6.21}$$

Thus, plotting the pH as a function of

$$\log \frac{OD_{obs} - OD_{AH}}{OD_A - OD_{obs}}$$

will yield a straight line which intercepts the y-axis at a pH equal to the pK.

6.3 Determination of the pK of 4-Methyl-2-Nitrophenol

Methylnitrophenol such as *p*-nitrophenol is a weak acid similar in its ionization behavior to phenol (Figure 6.1). It has the advantage of being colorless in its conjugate acid form and yellow in its conjugate base form (Figure 6.2).

6.3.1 Experimental procedure

Prepare a solution (2 ml) of methylnitrophenol in 10 mM Tris, pH 12. Add 10 μl of this solution to 1 ml of 10 mM Tris buffer (pH 2–11).

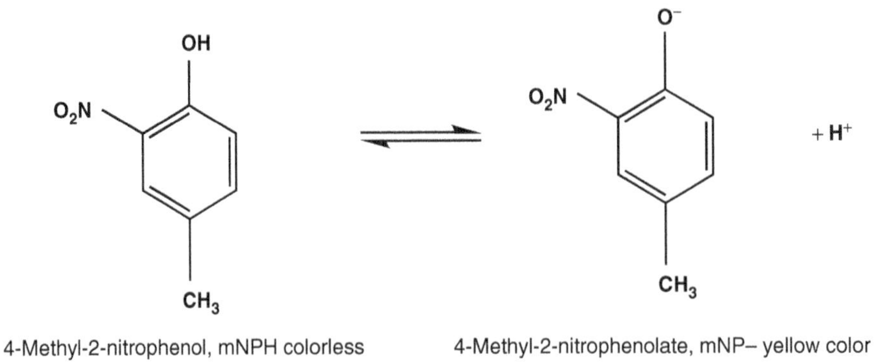

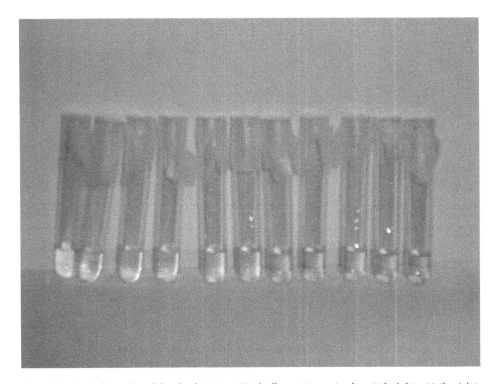

4-Methyl-2-nitrophenol, mNPH colorless   4-Methyl-2-nitrophenolate, mNP– yellow color

Figure 6.1   Ionization of 4-methyl-2-nitrophenol to 4-methyl-2-nitrophenolate.

Figure 6.2   Methylnitrophenol dissolved in 10 mM Tris buffer at pHs ranging from 2 (far left) to 12 (far right). The colorless protonated methylnitrophenol turns yellow when the pH increases. Reproduced in Color plate 6.2.

1   After each addition, plot the absorption spectrum of the solution from 300 to 500 nm. You will have 10 spectra characterizing absorption of methylnitrophenol at the different pHs (2–11). What do you notice about the absorption peak position? Is there any correlation between the absorption peak position and the color of the methylnitrophenol solution? Do you observe an isobestic point? If yes, explain its meaning.

2   Plot a graph showing the peak position of absorption spectrum of methylnitrophenol as a function of the pH. Determine the value of pK.

3   Plot a graph showing the optical density variation with pH at the peaks of the protonated (365 nm) and deprotonated (435 nm) forms of methylnitrophenol. What is the value of pK?

4   Plot, at 350 nm, a graph of the pH as a function of

$$\log \frac{OD_{obs} - OD_{AH}}{OD_A - OD_{obs}}.$$

What is the value of pK? What other information you can gain from the graph? Choose two or three other wavelengths and plot the graph. Do you obtain the same pK for all the graphs? Do you obtain the same pK with the three methods asked in 1, 2, and 3.

6.3.2 Solution

Color modification of the dye with the pH is accompanied by a shift in the absorption peak from short (365 nm) to higher wavelengths (435 nm). Figure 6.3 displays absorption

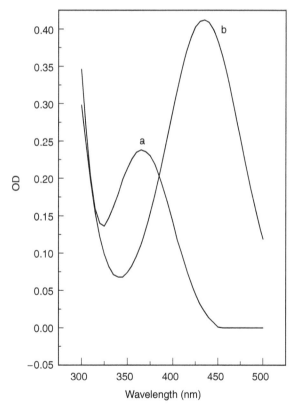

Figure 6.3 Absorption spectra of methylnitrophenol in Tris 10 mM, at pH 2 (a) and pH 11 (b). An isobestic point is observed at 385 nm.

spectra of methylnitrophenol at pH 2 and 11. An isobestic point is observed at 385 nm. At this wavelength, both forms of methylnitrophenol display equal absorption, which makes it difficult to study any of the two forms at this wavelength. At the isobestic point, the two absorbing forms of methylnitrophenol are in equilibrium, and their relative proportions are dependent on the concentration of hydrogen ions. For this reason, decreasing or increasing the pH will modify the concentration of the hydrogen ion, thereby displacing the equilibrium to the basic or to the acid form. In the presence of only two species, protonated and deprotonated, all the absorption spectra plotted at the different pHs, from 2 to 12, share a common isobestic wavelength equal to 385 nm (spectra not shown).

The optical density at the peak of deprotonated form is higher than that of the protonated form. Also, both deprotonated and protonated forms absorb at all wavelengths except at wavelengths higher than 450 nm where only the deprotonated form absorbs. Thus, experiments performed at wavelengths equal to or higher than 450 nm will be specific to the deprotonated form of methylnitrophenol.

Figure 6.4 shows the peak position of the methylnitrophenol absorbance spectrum at different pHs. Displacement of the peak position with the pH follows a sigmoidal curve, pK_a, determined at the inflection point of the curve is 6. This point corresponds to the maximum of the first derivative spectrum of the sigmoid curve and to the zero-crossing

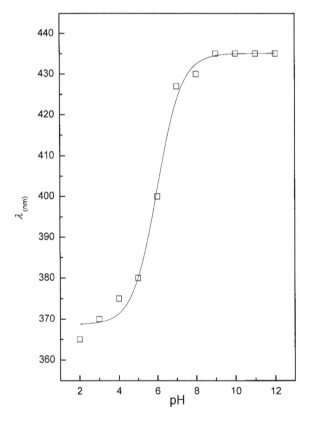

Figure 6.4 Position of the absorption peak of methylnitrophenol with pH.

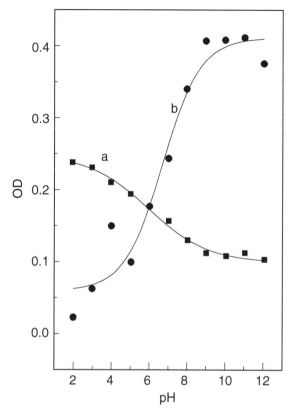

Figure 6.5 Variation in optical density with pH at the peaks of protonated (365 nm) (curve a) and deprotonated (435 nm) (curve b) forms of methylnitrophenol.

(the point where the signal crosses the x-axis going from either positive to negative or vice versa) in the second derivative.

Figure 6.5 shows the variation in optical density with pH at the peaks of the protonated (365 nm) and deprotonated (435 nm) forms of methylnitrophenol

The inflection point corresponding to pK is identical for both curves and is 6.2. Figure 6.6 shows the graph of the pH as a function of

$$\log \frac{OD_{obs} - OD_{AH}}{OD_{A} - OD_{obs}}$$

at 350 nm.

One can see that the pK is 6.2. This value is close to that (6) found for the pK with the two previous methods (peak variation and optical density variation with the pH). Therefore, the three methods are coherent and can be used to determine the pK of a dye.

Figure 6.6 shows clearly that it is possible to apply Equation (6.21) at pH 4–7. In this pH range, we obtain a linear plot according to the equation used. At pHs lower than 4 and higher than 7, Equation (6.21) can no longer be applied. Each dye has a specific pH application domain.

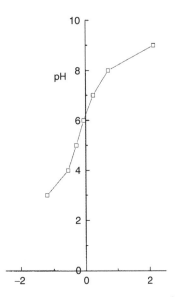

Figure 6.6 Graph of the pH of methylnitrophenol as a function of $\log \dfrac{OD_{obs}-OD_{AH}}{OD_A-OD_{obs}}$ at 350 nm. The pK is found to be equal to 6.2.

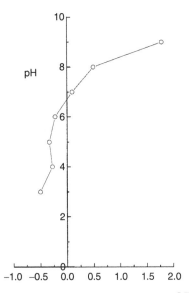

Figure 6.7 Graph of the pH of methylnitrophenol as a function of $\log \dfrac{OD_{obs}-OD_{AH}}{OD_A-OD_{obs}}$ at 410 nm. The pK is found to be equal to 6.7.

Plotting the same graph at wavelengths equal to or higher than 400 nm yields a pK equal to 6.7. At these wavelengths, the pH domain in which Equation (6.21) can be applied is restricted to two or three pH (6–8) (Figure 6.7). This result indicates that at wavelengths where absorption is dominated mainly by one species, the pK determined by this method would be more dependent on the form of the dye.

Finally, pK determination is also dependent on the ionic strength of the solvent used. In the literature, the pK_a of 4-methyl-2-nitrophenol is calculated to be equal to 7.11. Instructors and students can refer to Aptula *et al.* (2002).

Reference

Aptula, A.O., Netzeva, T.I., Valkova, I.V. *et al.* (2002). Multivariate discrimination between modes of toxic action of phenols. *Quantitative Structure–Activity Relationships* **21**, 12–22.

Chapter 7
Fluorescence Spectroscopy Principles

7.1 Jablonski Diagram or Diagram of Electronic Transitions

Absorption of light (photons) by a population of molecules induces electrons passage from the singlet ground electronic level S_0 to an excited state S_n $(n > 1)$. An excited molecule will return to the ground state S_0 following two successive steps:

1 The molecule at S_n returns to the lowest excited state S_1 by dissipating a part of its energy in the surrounding environment. This phenomenon is usually called internal conversion.

2 From the excited state S_1, the molecule will reach the ground state S_0 via different competitive processes:
   - Emission of a photon (fluorescence) with a radiative rate constant k_r.
   - Part of the absorbed energy is dissipated in the medium as heat. This type of energy is nonradiative and occurs with a rate constant k_i.
   - Excited molecules can release some of their energy to molecules located nearby. This energy transfer occurs with a rate constant k_q (collisional quenching), or with a rate constant k_t (energy transfer at distance).
   - A transient passage occurs to the excited triplet state T_1 of an energy lower than S_1 with a rate constant k_{isc} (inter-system crossing). For each excited state S, there is an excited state T of lower energy. The triplet state is an excited state and so is energetically unstable.

Therefore, de-excitation of the molecule from the triplet state occurs via competitive processes similar to those described for the de-excitation of the excited singlet state S_1:

1 Emission of a photon with a rate constant k_p. This phenomenon is called phosphorescence.

2 Dissipation of nonradiative energy with a rate constant k_i'.

3 Transfer of energy to another molecule at distance (rate constant k_t') or by collision (rate constant k_q') (Figure 7.1).

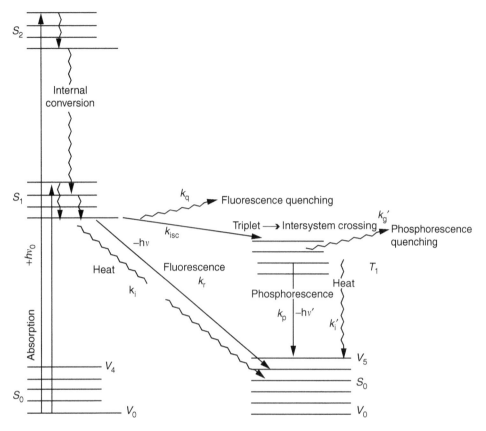

Figure 7.1 Jablonski (electronic transitions) diagram. Adapted from Jablonski, A. (1935). *Zeitschrift fur Physik*, **94**, 38–64.

A chromophore that emits a photon is called a fluorophore. Many chromophores do not necessarily fluoresce. In this case, energy absorbed is dissipated within the environment as thermal energy, collisional energy as the result of permanent collisions with the solvent molecules, and also energy transfer to other molecules. For example, heme does absorb light, but it does not fluoresce. The absence of fluorescence is the result of total energy transfer from the porphyrin ring to iron.

The Jablonski diagram is also called the electronic transitions diagram, since electrons of chromophores and/or fluorophores are responsible for the different described transitions.

The terminologies singlet and triplet state are the result of the spin quantum number s. Electron spinning around its own axis generates a local small magnet with a spin s. Singlet and triplet states depend on the quantum number of spin of the electron (s). According to the Pauli Exclusion Principle, two electrons within a defined orbital cannot have four equal quantum numbers and so differ in spin number. Two spin numbers are attributed to an electron, $+\frac{1}{2}$ and $-\frac{1}{2}$. Therefore, two electrons that belong to the same orbital will have opposite spins.

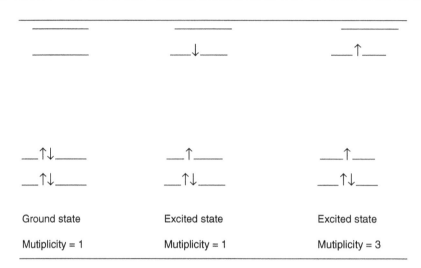

Figure 7.2 Spin configurations of the singlet and triplet states.

A parameter called the multiplicity (M) is defined as

$$M = |s_1 + s_2| + 1 \tag{7.1}$$

When the spins are parallel, i.e., $M = 1$, we have a singlet state S. When the spins are anti-parallel, i.e., $M = 3$, we have a triplet state T (Figure 7.2).

Upon light absorption, electrons are under the effect of electric and magnetic components. The contributions of these two components in the transition process could be defined as

$$F = (eE) + (evH/c) \tag{7.2}$$

where $e, c, v, E,$ and H are, respectively, the electron charge, light velocity, speed of rotation of the electron on itself, and electric and magnetic components of the light wave. The products (Ee) and (evH/c) are, respectively, the contributions of the electric and magnetic fields to the absorption phenomenon.

The speed of electron rotation on itself is very weak compared to the light velocity, and so the magnetic contribution to absorption is negligible compared to the electric contribution. Therefore, upon absorption, electronic transitions result from the interaction between electrons and electric field. For this reason, during absorption, a displaced electron preserves the same spin orientation. This is why only the $S_0 \rightarrow S_n$ transitions are allowed, and the $S_0 \rightarrow T_n$ transitions are forbidden.

In the ground state, all molecules except oxygen are in the singlet form. Oxygen is in the triplet state, and when excited, it reaches the singlet state. The singlet-excited state of oxygen is destructive for cells having tumors. Such interactions are exploited for biomedical applications known as photodynamic therapy. Compounds such as porphyrins are localized in target cells and tissues. Upon light activation, energy transfer occurs from porphyrins to oxygen molecules, thereby inducing a triplet $\rightarrow$ singlet transition within the oxygen state. Cells having tumor attract excited oxygen molecules and are then destroyed by these same oxygen molecules.

7.2   Fluorescence Spectral Properties

7.2.1   General features

Absorption allows a chromophore to reach an excited state, and so photon absorption induces excitation.

The absorption energy is higher than the emission energy. In fact, the total energy absorbed by the molecule is released in the medium in different ways, such as photon emission. Thus, the energy of the emitted photons is lower than the energy of the absorbed photons.

Many phenomena other than fluorescence emission contribute to fluorophore de-excitation. These other alternatives to fluorescence are radiationless loss, phosphorescence, photo-oxidation, and energy transfer. Thus, the weaker the competitive phenomena, the higher the de-excitation via fluorescence.

Emission occurs from the excited state S_1, independently of the excitation wavelength. Therefore, the emission energy would be independent of the excitation wavelength. The fluorescence energy is higher than that of phosphorescence.

Absorption and fluorescence do not require any spin reorientation. However, intersystem crossing and phosphorescence require a spin reorientation. Therefore, absorption and fluorescence are much faster than phosphorescence. Absorption occurs within a time equal to 10^{-15} s, and the fluorescence lifetime goes from 10^{-9} to 10^{-12} s. Phosphorescence is a long transition that can last from milliseconds to seconds, minutes, or even hours.

The energy of electronic transition is equal to the energy difference between the starting energy level and the final level. Therefore, the transition energy E (J mol^{-1}) is:

$$E = h\nu = hc/\lambda \tag{7.3}$$

where h is the Planck constant ($h = 6.63 \times 10^{-34}$ J$\cdot$s), ν is the light frequency (s^{-1} or Hertz, Hz), c is the light velocity, and λ is the wavelength (nm). Therefore, each transition occurs with a specific energy and so at a specific and single wavelength. However, as explained in the first chapter, spectra are observed, and not single lines, as the results of the contribution of rotational and vibrational levels to absorption and de-excitation energy. The fluorescence spectrum generated shows a maximum corresponding to the emission transition.

The absorption spectrum occurs from the ground state. Therefore, it will characterize the electronic distribution in this state. Fluorescence and phosphorescence occur from excited states, and so they are the mirrors of electronic distribution within the excited states, S for fluorescence and T for phosphorescence. Any modification of the electronic distribution in these states, such as in the presence of a charge transfer, will modify the corresponding spectrum. One such example is the reduction of cytochromes. The addition of an electron to the ground state, for example, modifies the electronic distribution within the molecule affecting the absorption spectrum.

Emission occurs from a population of n excited fluorophores with intensity I:

$$I = nE \tag{7.4}$$

The emission lifetime is within the picosecond-to-nanosecond range. Thus, emission is a very fast process, and so in order to observe fluorescence emission, the fluorophore should be excited continuously.

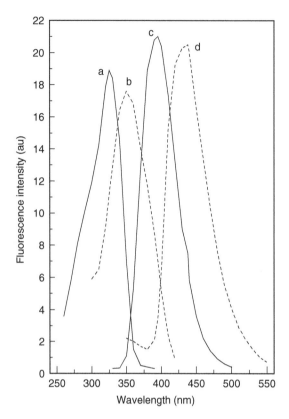

Figure 7.3 Fluorescence excitation (a and b) and emission (c and d) spectra of the 7-amino-4-methylcoumarin–DEVD complex (spectra a and c) and of 7-amino-4-methylcoumarin, obtained after autodissociation of the 7-amino-4-methylcoumarin–DEVD complex (spectra b and d). DEVD = Asp–Glu–Val–Asp. 7-Amino-4-methylcoumarin–DEVD complex: (a) λ_{em} = 470 nm and λg_{max} = 323 nm. (c) λ_{ex} = 260 nm and λ_{max} = 393 nm. 7-Amino-4-methylcoumarin: (b) λ_{em} = 440 nm and λ_{max} = 350 nm. (d) λ_{ex} = 300 nm and λ_{max} = 438 nm.

Temperature variation induces modification of global and local motions of the fluorophore environment and of the fluorophore itself, modifying its fluorescence emission feature.

The intensity, position of the emission wavelength, and lifetime are some of the observables that will characterize a fluorophore. Each fluorophore has its own fluorescence properties and observables. These properties are intrinsic to the fluorophore and are modified with the environment. A fluorescence spectrum is the plot of the fluorescence intensity as a function of wavelength (Figure 7.3).

Since we have to excite the sample and record the emitted intensity at different wavelengths, the layout of a fluorometer consists of an excitation source (a lamp or a laser), an excitation monochromator or a filter (if a lamp is used as the source of excitation), a cuvette holder into which we can put the sample, an emission monochromator or a filter

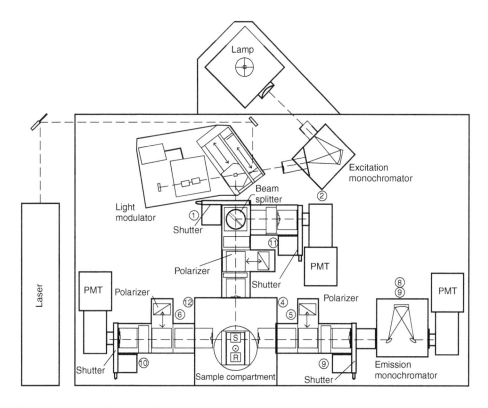

Figure 7.4 Layout of a spectrofluorometer. Courtesy of ISS Instruments.

(if we do not want to record the whole spectrum but just the fluorescence intensity), a photon detector, and a recorder (Figure 7.4).

When we want to record the fluorescence emission spectrum, the excitation wavelength is kept fixed, and the emission monochromator is run (Figures 7.3c and d). The excitation spectrum is obtained by running the excitation monochromator at a fixed emission wavelength (Figures 7.3a and b).

7.2.2 Stokes shift

The energy absorbed by a fluorophore is more important than the energy of an emitted photon. Referring to Equation (7.3), the absorption energy is

$$E_a = hc/\lambda_a \tag{7.5}$$

and emission energy is

$$E_{em} = hc/\lambda_{em} \tag{7.6}$$

Since

$$E_{em} < E_a \tag{7.7}$$

We have

$$\lambda_{em} > \lambda_a \tag{7.8}$$

λ_a and λ_{em} are absorption and emission spectra peaks.

Thus, the emission spectrum has its maximum shifted to longer wavelengths compared to the maximum of absorption spectrum (Figure 7.3). Sir George Stokes observed this shift for the first time in 1852 and since this time it is called the Stokes shift.

When absorption and/or the emission spectra of a fluorophore possess two or more bands, the Stokes shift is equal to the difference that separates the two most intense bands of the two spectra.

7.2.3   Relationship between the emission spectrum and excitation wavelength

Emission occurs from the excited state S_1, and so is in principle independent from the excitation wavelength.

Since not all the molecules present at the excited states will participate in the fluorescence process, a quantum yield Φ_F exists:

$$\Phi_F = \frac{I_F}{I_A} \tag{7.9}$$

$\Phi_F < 1$, since

$$I_A = I_0 - I_F \tag{7.10}$$

and

$$I_T = I_0 10^{-\varepsilon cl} = I_0 e^{-2.3\varepsilon cl} \tag{7.11}$$

$$I_A = I_0 - I_0 e^{-2.3\varepsilon cl} = I_0(1 - e^{-2.3\varepsilon cl}) \tag{7.12}$$

At very low optical densities ($\ll 0.05$), Equation (7.12) can be written as

$$I_0(l - (l - 2.3\varepsilon_{(\lambda)}cl)) = 2.3 I_0 \varepsilon_{(\lambda)} cl \tag{7.13}$$

Thus,

$$I_{F(\lambda)} = 2.3 I_0 \varepsilon_{(\lambda)} cl \Phi_F \tag{7.14}$$

Equation (7.14) describes the relationship that exists between fluorescence intensity at a precise emission wavelength and intensity of incident light (I_0), fluorophore quantum yield, and its optical density at the excitation wavelength.

I_0 is the intensity of the incident beam that will excite the sample. Thus, excitation with a laser will allow yield a better fluorescence intensity than if a lamp were used for excitation. The fluorescence intensity is proportional to the quantum yield, i.e., a high quantum yield leads automatically to a high fluorescence intensity.

The fluorescence intensity is proportional to the optical density. This means that it is proportional to εc and l. Since ε is wavelength-dependent, the fluorescence intensity is also wavelength-dependent.

7.2.4 Inner filter effect

The fluorescence intensity is linearly proportional to the optical density (Equation (7.14)). However, let us take a usual fluorometer where the emission beam is recorded perpendicularly to the excitation beam. In general, emission is detected at the cuvette center. Thus, the incident beam will go through half the cuvette path length before emission recording. Also, the emission beam induced at the center of the cuvette should go through half of the cuvette path length before being detected. Therefore, the recorded emission intensity will be underestimated as the result of the optical density at both excitation and emission wavelengths. This auto-absorption by the solution is called the inner filter effect. It can modify the whole fluorescence spectrum inducing a shift in the emission maximum and a decrease in the fluorescence intensities.

The fluorescence intensity for the inner filter effect can be corrected using Equation (7.15) (Lakowicz, 1999; Albani, 2004):

$$F_{corr} = F_{rec} * 10^{[OD(em)+OD(ex)]/2} \tag{7.15}$$

7.2.5 Fluorescence excitation spectrum

The fluorescence excitation spectrum characterizes the electron distribution of the molecule in the ground state. Excitation is, in principle and for a pure molecule, equivalent to absorption. The fluorescence excitation spectrum is obtained by fixing the emission wavelength and by running the excitation monochromator (Figure 7.3).

The excitation spectrum is technically perturbed by two problems: the light intensity of the excitation lamp, which varies with the wavelength, and the intensity upon detection, which is also wavelength-dependent. Corrections can be performed using rhodamine B, dissolved in glycerol, as reference. In fact, radiation from rhodamine is proportional to the excitation intensity independently of the excitation wavelengths. Therefore, excitation of rhodamine will yield a fluorescence excitation spectrum that characterizes excitation lamp spectrum. In order to obtain the real fluorescence excitation spectrum of the studied fluorophore, the recorded excitation spectrum will be divided by the excitation spectrum obtained from rhodamine. This procedure is done automatically within the fluorometer.

In general, when one wants to determine if global and/or local structural modifications have occurred within a protein, circular dichroism experiments are performed. Also, one can record the fluorescence excitation spectrum of the protein. If perturbations occur within the protein, one should observe excitation spectra that differ from one state to another. One should not forget to correct the recorded spectra for the inner filter effect.

7.2.6 Mirror–image rule

The emission spectrum of a fluorophore is the image of its absorption spectrum when the probability of $S_1 \rightarrow S_0$ transition is identical to that of $S_0 \rightarrow S_1$ transition. If, however, excitation of the fluorophore leads to an $S_0 \rightarrow S_n$ transition, with $n > 1$, internal relaxation will occur so that molecules reach the first excited singlet state before emission.

This will induce an emission transition different from the absorption one. The mirror–image relationship is generally observed when the interaction of the fluorophore excited state with the solvent is weak.

7.2.7   Fluorescence lifetime

7.2.7.1   Definition of fluorescence lifetime

After excitation, molecules remain in the excited state for a short time before returning to the ground state. The excited-state lifetime is equal to the mean time during which molecules remain in the excited state. This time is considered as the fluorescence lifetime. This time ranges from the nanoseconds (10^{-9} s) to picoseconds (10^{-12} s).

The fluorescence lifetime τ_f is:

$$\tau_f = 1/k = 1/(k_r + k_{isc} + k_i) \tag{7.16}$$

The mathematical definition of fluorescence lifetime arises from the fact that nonradiative and radiative processes participate in the fluorophore deexcitation.

The radiative lifetime τ_r is $1/k_r$. It is the real emission lifetime of a photon that should be measured independently of the other processes that deactivate the molecule. However, since these processes occur in parallel to the radiative process, it appears impossible to eliminate them during radiative lifetime measurements. Therefore, we will measure a time characteristic of all deexcitation processes. This time is called the fluorescence lifetime and is lower than the radiative lifetime. A fluorophore can have one or several fluorescence lifetimes; in this case, we can determine the fractional contribution of each lifetime and calculate the mean fluorescence lifetime τ_0 or $\langle \tau \rangle$:

$$\tau_0 = \sum f_i \tau_i \tag{7.17}$$

and

$$f_i = \beta_i \tau_i \Big/ \sum \beta_i \tau_i \tag{7.18}$$

where β_i is the pre-exponential term, τ_i is the fluorescence lifetime, and f_i is the fractional intensity.

A fluorophore can have one or several fluorescence lifetimes, depending on several factors:

1 A ground-state heterogeneity resulting from the presence of equilibrium between different conformers. Each conformer presents a specific fluorescence lifetime.
2 Different nonrelaxing states of the Trp emission. These relaxed states may arise from one single Trp residue.
3 Internal protein motion.
4 The presence in the studied macromolecule, for example a protein, of two or several tryptophans each having different microenvironments and so different emissions.

The nature of the environment and of its interaction with the fluorophore can affect all the fluorescence parameters.

Also, it is important to mention that in many cases, fluorescence lifetimes are dependent on the structure of the fluorophore itself (Albani, 2007).

In general, fluorescence intensity decreases exponentially with time. Let us consider an N_0 fluorophores population having reached the excited state. The velocity of decrease in this population with time t is

$$-\frac{dN(t)}{dt} = (k_r + k')N(t) \tag{7.19}$$

where k_r is the radiative rate constant, k' the sum of rate constant of all other competing processes, and $N(t)$ the population at the excited state at time t.

$$\frac{dN(t)}{N(t)} = -(k_r + k')dt \tag{7.20}$$

$$\log N = -(k_r + k')t + \text{constant} \tag{7.21}$$

At $t = 0, N = N_0$, and $\log N = \log N_0$

$$\log N = -(k_r + k')t + \log N_0 \tag{7.22}$$

$$\log N/N_0 = -(k_r + k')t \tag{7.23}$$

$$N/N_0 = e^{-(k_r+k')t} \tag{7.24}$$

$$N = N_0 e^{-(k_r+k')t} \tag{7.25}$$

$$\tau = 1/(k_r + k') \tag{7.26}$$

$$N = N_0 e^{-t/\tau} \tag{7.27}$$

When $t = \tau$:

$$N = N_0 e^{-1} = N_0/e \tag{7.28}$$

Thus, we can define the fluorescence lifetime as the time required for the excited-state population to be reduced by $1/e$ of its initial population, immediately after excitation.

7.2.7.2 Fluorescence lifetime measurement

Three techniques are actually available for measuring the fluorescence lifetime: Strobe, Time Correlated Single Photon Counting (TCSPC), and multifrequency and cross-correlation spectroscopy. Strobe and TCSPC are based on measurement in the time domain, while multifrequency and cross-correlation spectroscopy measure fluorescence lifetimes in the frequency domain. The time domain allows direct observation of fluorescence decay, while the frequency domain is a more indirect approach in which the information regarding the fluorescence decay is implicit.

In the time-correlated single-photon counting (TCSPC) technique, the sample is excited with a pulsed light source. The light source, optics, and detector are adjusted so that, for a given sample, no more than one photon is detected. When the source is pulsed, a timer is started. When a photon reaches the detector, the time is measured. Over the course of the

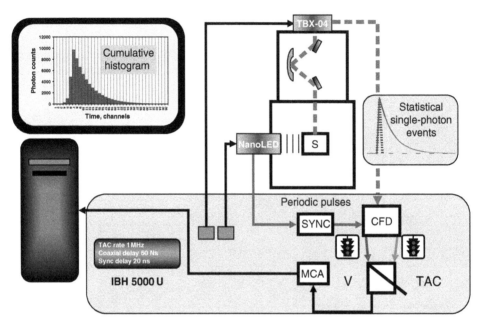

Figure 7.5 Time-correlated single-photon counting instrument principle. Courtesy of Jobin Yvon.

experiment, the fluorescence decay curve is constructed by measuring the "photon events," or accumulated counts, vs. time (Figure 7.5) (Badea and Brand 1979).

In the strobe or pulse sampling technique, the sample is excited with a pulsed light source. The intensity of the fluorescence emission is measured in a very narrow time window on each pulse and saved on the computer. The time window is moved after each pulse. When data have been sampled over the appropriate range of time, a decay curve of emission intensity vs. time can be constructed.

The name "strobe technique" comes about because the photomultiplier PMT is gated – or strobed – by a voltage pulse that is synchronized with the pulsed light source. The strobe has the effect of "turning on" the PMT and measuring the emission intensity over a very short time window (Figure 7.6) (Bennett 1960; James *et al.* 1992).

In frequency-domain instruments, the sample is excited by a light source whose intensity is sinusoidally modulated (w). Emitted light will also be sinusoidally modulated at the same frequency of the excited light but will be delayed by a phase angle, ϕ, due to the finite persistence of the excited state. The phase angle, ϕ, is equal to ($w\theta$), where θ is the phase delay time. The intensity of emitted light is lower than that of excitation light. The fluorescence lifetime of the sample can be calculated from the phase delay and/or from light demodulation. A long fluorescence lifetime yields a highly delayed emission and a large phase shift between excitation and emission lights (Figure 7.7) (Spencer 1970).

A glycogen solution placed in the emission compartment will scatter light and is used as reference ($\tau_f = 0$) to determine the phase delay and fluorescence demodulation. For each measurement, the reference intensity is adjusted so that it is equivalent to the intensity of the fluorescence signal of the sample. Phases and modulations of the fluorescence and scattered light are obtained relative to the reference photomultiplier or instrumental

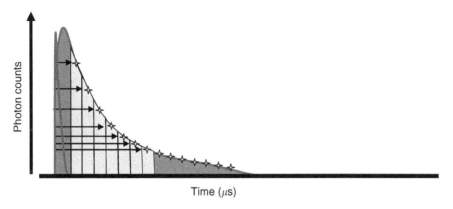

Figure 7.6 Schematic profile of the fluorescence lifetime measured by the Strobe method. Courtesy of Jobin Yvon.

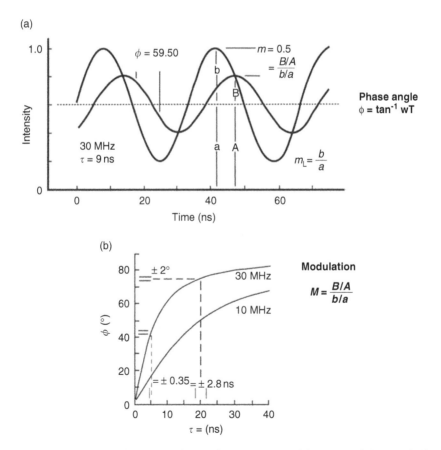

Figure 7.7 Determination of fluorescence lifetime from intensity modulation (a) and phase angle (b).

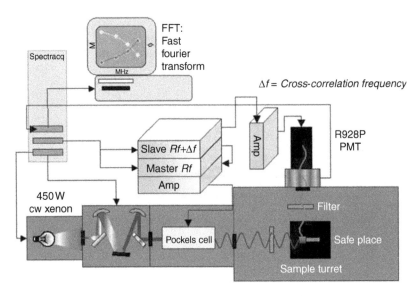

Figure 7.8 Multifrequency cross-correlation fluorometer. Courtesy of Jobin Yvon.

(internal) reference signal. In fact, two identical detection electronic systems are used to analyze the outputs of the reference and sample photomultipliers. Each channel consists of an alternative and continuous currents (*AC* and *DC*, respectively).

$$m = (AC/DC)_{EM}/(AC/DC)_{EX} = (B/A)/(b/a) \tag{7.29}$$

where "EM" and "EX" refer to fluorescence and scattered light. Lifetimes can be obtained by the phase method, denoted by τ_P, or by modulation, denoted by τ_M.

A mono-exponential decrease in fluorescence (one lifetime) yields $\tau_P = \tau_M$. A multi-exponential decrease gives $\tau_P < \tau_M$.

Multifrequency is achieved by using frequency modulators. The modulator currently used is an electro-optic modulator that is crossed transversely by an electric current (Pockels cell) (Figure 7.8). In the phase detection system, high-frequency signals are used. In the cross-correlation technique, the high-frequency signals are converted to low-frequency signals. Excitation light is first modulated and then sent to the cell compartment equipped with a revolving turret so that it can excite the studied sample, then the reference, which is a fluorophore of known fluorescence lifetime. This fluorophore replaces glycogen solution.

A part of the excitation light is sent toward a reference photomultiplier so that the intensity and phase of the excitation signal can be recorded. A frequency synthesizer connected to the Pockels cell gives the desired frequency. Rather than to apply a continuous current on the detection and reference photomultipliers, the two photomultipliers are submitted to a sinusoidal electric current $R(t)$ of frequency $(w + \Delta w)$ as close as possible to the frequency w of the fluorescence light. Δw is the cross-correlation frequency. The signal recorded by the detection photomultiplier is thus the product of the emitted light $F(t)$ and of $R(t)$. Therefore, the resulting signal evolves with a weak frequency Δw possessing all information "contained" in the fluorescent light. One measures the evolution of the value $\langle R(t) * F(t) \rangle$ of the current delivered by the photomultiplier. This method, called cross-correlation, compares the excitation and emission signals. This double procedure considerably increases the signal-to-noise ratio.

For more details on the three different techniques, readers can refer to the following references: Lakowicz (1999), Valeur (2002), and Albani (2004).

7.2.8 Fluorescence quantum yield

Molecules in the fundamental state absorb light with an intensity equal to I and reach an excited state S_n. Then, different competitive processes, including fluorescence, will compete with each other to de-excite the molecule. The rate constant (k) of the excited state is the sum of the kinetic constants of the competitive processes:

$$k = k_r + k_{isc} + k_i \tag{7.30}$$

The fluorescence quantum yield Φ_F is the number of photons emitted by the radiative way over that absorbed by the molecule:

$$\Phi_F = \frac{\text{emitted photons}}{\text{absorbed photons}} = \frac{k_r}{k_r + k_i + k_{isc}} \tag{7.31}$$

The fluorescence quantum yield of a molecule is obtained by comparing the fluorescence intensity of the molecule with that of a reference molecule with a known quantum yield:

$$\Phi_2 = \frac{OD_1 * \Sigma F_2}{OD_2 * \Sigma F_1} \Phi_1 \tag{7.32}$$

where F_2 is the fluorescence intensity of the molecule of unknown quantum yield Φ_2, and F_1 is the fluorescence intensity of the reference with quantum yield Φ_1.

Therefore, in order to determine fluorophore quantum yield, one needs to measure the optical densities of the fluorophore and of the reference at the excitation wavelength, and to calculate for each of them the sum of their fluorescence intensities along their fluorescence emission spectra.

In proteins, the quantum yield from the dominant fluorescent amino acid is calculated. For example, in a protein in which Trp residues are the main emitters, the quantum yield is determined by comparison with a solution of free L-Trp ($\Phi_F = 0.14$ at 20°C and at $\lambda_{ex} = 295$ nm). Also, quinine sulfate dissolved in 0.1 M sulfuric acid is commonly used as a reference or standard molecule. Its quantum yield is 0.55 with an absorption and emission maxima at 340 and 445 nm, respectively.

Finally, one should remember that the standard and the molecule to be analyzed should be studied under the same conditions of temperature and solvent viscosity. Also, it is always better to work at low optical densities in order to avoid corrections for the inner filter effect.

In a multitryptophan protein, the fluorescence quantum yield of tryptophans can be additive or not. In the first case, there is no interference between Trp residues. In the second case, energy transfer between tryptophan residues can influence the quantum yield of each of them.

In order to measure quantum yields of an extrinsic fluorophore bound to a protein and which emits at longer wavelengths than in the UV, standards such as 3,3'-diethylthiacarbocyanine iodide (DTC) in methanol ($\Phi_F = 0.048$) and rhodamine 101

Table 7.1 Fluorescence observables of tryprophan in solvents of different pHs

	Φ_F	λ_{max} (nm)	τ_0 (ns)
Basic	0.41	365	9
Neutral	0.14	351	3.1
Acid	0.04	344	0.7

in ethanol ($\Phi_F = 0.92$) or any other dyes can be used. Quantum yield is calculated according to Equation (7.33):

$$\frac{\Phi_S}{\Phi_R} = \frac{A_S}{A_R} \times \frac{OD_R}{OD_S} \times \frac{n_S^2}{n_R^2} \tag{7.33}$$

where, Φ_S and Φ_R are the fluorescence quantum yields of the sample and reference, respectively. A_S and A_R are areas under the fluorescence spectra of the sample and the reference, respectively; $(OD)_S$ and $(OD)_R$ are the respective optical densities of the sample and the reference solution at the excitation wavelength; and n_S and n_R are the values of the refractive index for the respective solvents used for the sample and reference (Barik *et al.* 2003).

Table 7.1 shows the fluorescence lifetime, quantum yield, and position of the emission maximum of tryptophan in basic, neutral, and acidic solvents. One can see that the three parameters are not the same in the three media. The different types of protonation explain this variation.

7.2.9 Fluorescence and light diffusion

There are two types of light diffusion: Rayleigh and Raman. Energies of diffused and excited photons are equal. Thus, Rayleigh diffusion and excitation occurs at the same wavelengths.

Usually, one should not start recording the emission spectrum before or at the excitation wavelength in order to avoid recording Rayleigh diffusion. Raman diffusion is observed at a wavelength higher than the excitation wavelength. Diffused photons have a quantum energy lower than that of the excitation photons. Therefore, the intensity of the Raman peak is lower than that of the Rayleigh peak. Raman diffusion can perturb emission spectrum, and so one should substract Raman from the emission spectrum. In general, the higher the fluorescence intensity, the lower the Raman peak (Figure 7.9). In an aqueous medium, the O—H bonds are responsible for the Raman spectrum. The position of the Raman peak is dependent on the excitation wavelength. It is possible to calculate the position of the Raman peak in an aqueous medium using the following equation:

$$1/\lambda_{ram} = 1/\lambda_{ex} - 0.00034 \tag{7.34}$$

The wavelengths are in nanometers.

7.3 Fluorophore Structures and Properties

Fluorophores, small molecules that can be part of a molecule (intrinsic fluorophores) or added to it (extrinsic fluorophores), can be found in different cells, and so they can be

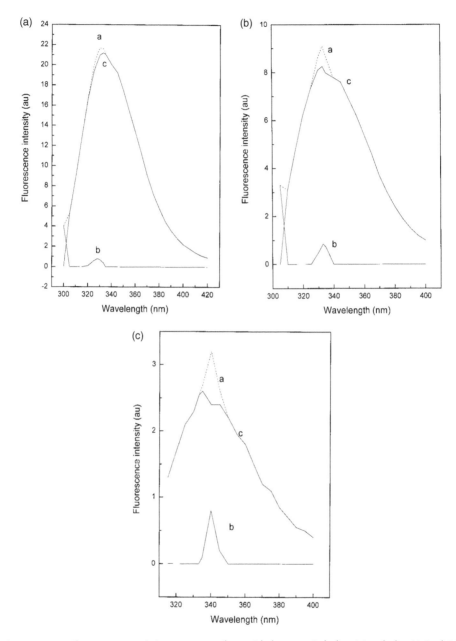

Figure 7.9 (a) Fluorescence emission spectrum of α_1-acid glycoprotein before (a) and after (c) Rayleigh and Raman substraction of the buffer (spectrum b). Raman peak makes up 3.75% of the peak in spectrum (a). $\lambda_{ex} = 295$ nm. (b) Fluorescence emission spectrum of α_1-acid glycoprotein before (a) and after (c) Rayleigh and Raman substraction of the buffer (spectrum b). Raman peak makes up 9.34% of the peak in spectrum (a). $\lambda_{ex} = 300$ nm. (c) Fluorescence emission spectrum of α_1-acid glycoprotein before (a) and after (c) Raman substraction of the buffer (spectrum b). The Raman peak makes up 25% of the peak in spectrum (a). $\lambda_{ex} = 305$ nm.

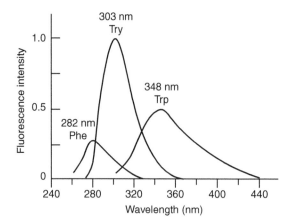

Figure 7.10   Fluorescence emission spectra of aromatic amino acids.

used as natural indicators to study the structure, dynamics, and metabolism of living cells. Their fluorescence properties are dependent on their structure and on the surrounding environment. Each fluorophore has its own specific fluorescence properties. Intrinsic fluorophores are aromatic amino acids and cofactors. Extrinsic fluorophores are those which can be linked covalently or not to macromolecules such as peptides, proteins, membranes, or DNA.

7.3.1   Aromatic amino acids

In proteins, tryptophan, tyrosine, and phenylalanine are responsible for the absorption and fluorescence of proteins in the UV (Figures 1.2 and 7.10).

Tryptophan fluorescence is very sensitive to the local environment. In an environment with a low polarity, tryptophan emits at a maximum of 320 nm. The peak position shifts to 355 nm in the presence of a polar environment. The loss of the protein tertiary structure (complete denaturation) induces a shift in tryptophan fluorescence to 355 nm.

Also, upon binding of a ligand to a protein, Trp observables (intensity, polarization, and lifetime) can be altered, and so one can follow this binding with Trp fluorescence. In proteins, tryptophan fluorescence dominates. Zero or weak tyrosine and phenylalanine fluorescence results from energy transfer to tryptophan and/or neighboring amino acids.

Burstein *et al.* (1973) classified tryptophan in proteins into three categories, according to the position of their fluorescence maximum (λ_{max}) and the bandwidth ($\Delta\lambda$) of their spectrum:

- Category 1: λ_{max} 330–332 nm and $\Delta\lambda$ 48–49 nm.
- Category 2: λ_{max} 340–342 nm and $\Delta\lambda$ 53–55 nm.
- Category 3: λ_{max} 350–353 nm and $\Delta\lambda$ 59–61 nm.

When a protein contains two classes of Trp residues, the recorded fluorescence emission spectrum is the result of each class contribution.

The absorption spectra for the aromatic amino acids, displayed in Figure 1.2, overlap at many wavelengths. Therefore, if we want to observe emission from the Trp residues alone,

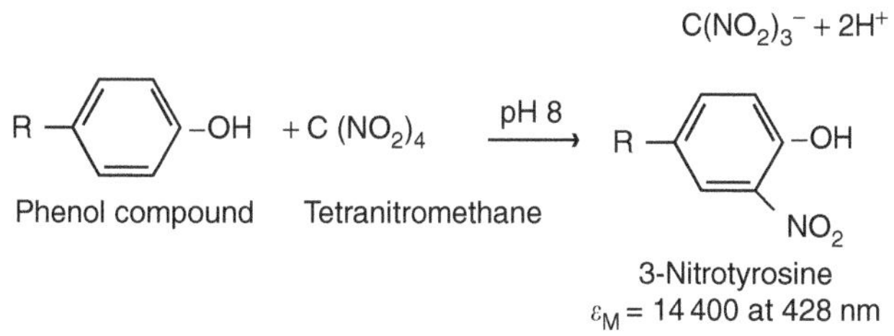

$$C(NO_2)_3^- + 2H^+$$

R—⟨benzene⟩—OH + C(NO$_2$)$_4$ $\xrightarrow{\text{pH 8}}$ R—⟨benzene⟩—OH

Phenol compound   Tetranitromethane

NO$_2$

3-Nitrotyrosine

$\varepsilon_M = 14\,400$ at 428 nm

Figure 7.11 Nitration of tyrosine with TNM. Reproduced with permission from Villette J.R., Helbecque, N., Albani, J.R., Sicard, P.J. and Bouquelet, S.J. (1993). *Biotechnology and Applied Biochemistry*, **17**, 205–216. © Portland Press Ltd.

excitation should occur at 295 nm and/or above. At these wavelengths, energy transfer from tyrosine to tryptophan does not take place, and the emission observed emanates from the tryptophan residues only.

Tyrosine is more fluorescent than tryptophan in solution, but when present in proteins, its fluorescence is weaker. This can be explained by the fact that the protein tertiary structure inhibits tryosine fluorescence. Also, energy transfer from tyrosines to tryptophan residues occurs in proteins inducing a total or important quenching of tyrosine fluorescence. This tyrosine → tryptophan energy transfer can be evidenced by nitration of tyrosine residues with tetranitromethane (TNM), a highly potent pulmonary carcinogen. Because TNM specifically nitrates tyrosine residues on proteins, the effects of TNM on the phosphorylation and dephosphorylation of tyrosine, and the subsequent effects on cell proliferation, can be investigated.

Nitration of phenolic compounds leads to the formation of 3-nitrotyrosine (molar extinction coefficient $= 14\,400$ at 428 nm) (Figure 7.11). The reaction is very specific for phenolic compounds. Chemical nitration of functionally important tyrosine residues by tetranitromethane has often been found to inactivate or alter the enzyme properties. It was only after the detection of *in vivo* nitrotyrosine formation under inflammatory conditions that the physiological aspects of nitrotyrosine metabolism came to light. Abundant production (1–120 μM) of nitrotyrosine has been recorded under a number of pathological conditions such as rheumatoid arthritis, liver transplantation, septic shock, and amyotrophic lateral sclerosis (Balabanli *et al.* 1999).

CGTases (EC 2.4.1.19) are bacterial enzymes that facilitate the biosynthesis of cyclodextrins from starch through intramolecular transglucosylation. The primary structures of most of these enzymes have been published, and the three-dimensional structure of *Bacillus circulans* CGTase has been established. Studies of transglucosylation molecular mechanism have indicated that amino acids such as histidine and tryptophan are implicated in such mechanisms. Nitration of CGTase with TNM induces a loss of enzyme activity, a decrease in enzyme affinity towards the β-CD copolymer, and a loss of tryptophan fluorescence (Villette *et al.* 1993).

The fluorescence emission maximum of CGTase is located at 338 nm, and its spectrum bandwidth is 55 nm (Figure 7.12a). Thus, both embedded and surface tryptophan residues contribute to protein fluorescence. Although CGTase contains many tyrosine residues, the absence of a shoulder or a peak at 303 nm (Figure 7.12b), when excitation is performed at 273 nm, suggests that tyrosine residues do not contribute to CGTase emission.

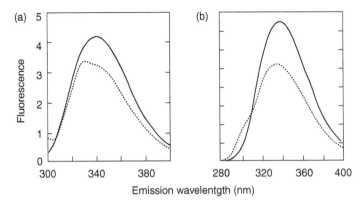

Figure 7.12 Fluorescence of CGTase obtained at λ_{ex} = 295 (a) and 273 nm (b). (——) : native enzyme, (- - - -) : 8 mM TNM modified protein. Source: Reproduced with permission from Villette J.R., Helbecque, N., Albani, J.R., Sicard, P.J. and Bouquelet, S.J. (1993). *Biotechnology and Applied Biochemistry, 17*, 205–216. © Portland Press Ltd.

The tryptophan fluorescence intensity of native CGTase is eightfold higher than that of a L-tryptophan solution with respect to the total amount of tryptophan. This suggests that CGTase supports energy transfer from tyrosyl to tryptophan residue(s).

The fluorescence intensity of CGTase tryptophan residues decreases (Figure 7.13) after treatment of the enzyme with increasing concentrations of TNM and purification of the modified enzyme on co-polymer. Tryptophan fluorescence intensity of the 8 mM-TNM-modified CGTase (0.03 μM^{-1} tryptophan) is similar to that of free L-tryptophan (0.034 μM^{-1}). The loss of tryptophan fluorescence observed during nitration may then be related to elimination of this energy transfer.

CGTase nitration induces an 11.5 nm shift in the fluorescence maximum to shorter wavelengths (λ_{max} = 326.5 nm instead of 338 nm for 8 mM TNM), suggesting a relative increase in the buried tryptophan fluorescence. Tyr → Trp energy transfer may then involve solvent-exposed residue(s).

After excitation of nitrated CGTase at 273 nm (Figure 7.12b), tyrosine fluorescence appears as a typical shoulder with a peak at 302 nm. This phenomenon, not observed in native enzyme, is consistent with elimination of Tyr → Trp energy transfer during enzyme nitration.

Circular dichroism spectra have been recorded on native and 8 mM TNM-treated CGTases in the 210–250 and 250–320 nm regions (data not shown) in order to investigate nitration effects on both the secondary protein structure and the conformational environment of aromatic residues.

The spectrum of native CGTase shows two typical peaks in the first region at 212 and 220 nm. After nitration with 8 mM TNM, the ellipticity decreases, and the 212 nm peak disappears. This result suggests a slight impairment in CGTase conformation during its nitration by TNM.

The 250–320 nm spectrum shows an important change around 280 nm for nitrated CGTase in comparison with that of native enzyme. This is in part due to tyrosine nitration. In particular, the dichroism observed around 300 nm in the differential spectrum (results

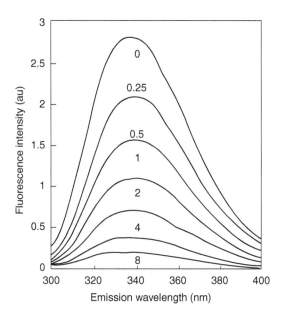

Figure 7.13 Tryptophan fluorescence spectra of CGTase after 18 h of nitration with 0, 0.25, 0.5, 1, 2, 4, and 8 mM TNM. λ_{ex} = 295 nm. Source: Villette J.R., Helbecque, N., Albani, J.R., Sicard, P.J. and Bouquelet, S.J. (1993). *Biotechnology and Applied Biochemistry*, **17**, 205–216. Reprinted with permission from Portland Press.

Table 7.2 Absorption and fluorescence properties of tryptophan, tyrosine, phenylalanine, and the Y-base

Fluorophore	Conditions	Absorption		Fluorescence			
		λ_{max} (nm)	$\varepsilon_{max}(\times 10^{-3})$ $(M^{-1}\,cm^{-1})$	λ_{max} (nm)	Φ_F	τ_f (ns)	Sensitivity
Trp	H_2O, pH 7	280	5.6	348	0.2	2.6	11
Tyr	H_2O, pH 7	274	1.4	303	0.1	3.6	2
Phe	H_2O, pH 7	257	0.2	282	0.04	6.4	0.08
Y-base	t-ARN^{Phe} From yeast	320	1.3	460	0.07	6.3	0.91

Reference: Cantor, C.R. and Schimmel, P.R. (1980). *Biophysical Chemistry*, W.H. Freeman, New York.

not shown) strongly suggests that the conformational environment of some tryptophan residues has been changed during nitration.

These results support a conformational impairment of the enzyme during nitration and would account for the loss of tryptophan fluorescence by the elimination of (Tyr → Trp) energy transfer (removal of the chromophores) and hence enzyme inactivation. Table 7.2 lists the principal absorption and fluorescence characteristics of the three aromatic amino acids within the proteins and of the Y-base.

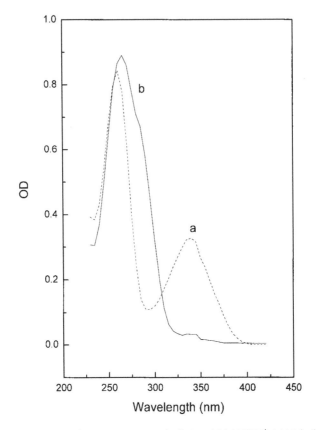

Figure 7.14 Absorption spectra for (a) NADH (pH 7 buffer) and (b) NADH$^+$ (pH 2 buffer).

7.3.2 Cofactors

NADH is highly fluorescent, with absorption and emission maxima located at 340 and 450 nm, respectively, while NAD$^+$ and NADH$^+$ are not fluorescent (Figures 7.14 and 7.15). Schauenstein *et al.* (1980) explained the spectral modification of NADH$^+$ by a higher percentage of the stacked conformation at a lower pH.

FMN and FAD absorb light in the visible at 450 nm and fluoresce at around 515 nm. The fluorescence lifetimes of FMN and FAD are 4.7 and 2.3 ns, respectively.

7.3.3 Extrinsinc fluorophores

7.3.3.1 Fluorescein and rhodamine

These bind covalently to lysines and cysteines of proteins, and absorb and fluoresce in the visible. The fluorescence lifetimes of fluorescein and rhodamine are around 4 ns, and their emission spectra are not sensitive to medium polarity.

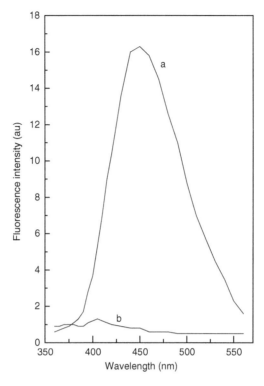

Figure 7.15   Fluorescence emission spectra of (a) NADH (pH 7) and of (b) NADH$^+$ (pH 2). $\lambda_{ex} = 340$ nm.

7.3.3.2   *Naphthalene sulfonate*

5-((((2 iodoacetyl)amino)ethyl)amino)naphthalene-1-sulfonic acid (IAEDANS or 1,5-IAEDANS) and DNS bind covalently to proteins, while l-anilino-8-naphthalene sulfonate (ANS) and 2-*p*-toluidinylnaphthalene-6-sulfonate (TNS) bind non-covalently to proteins and membranes. Dissolved in a polar medium such as water, TNS and ANS show a very weak fluorescence that increases with decreasing medium polarity. When bound to proteins or to membranes, their fluorescence increases, and their maximum shifts to the blue edge. The intensity increase and the shift to shorter wavelengths are increasingly important with the decrease in binding-site polarity. Table 7.3 lists the characteristics of several extrinsic probes.

7.3.3.3   *Nucleic bases*

Puric and pyrimidic bases show a weak fluorescence in aqueous medium, which increases by a factor of 10 in a medium of pH 2 and by a factor of 100 at 77 K. The quantum yield of the nucleic acids depends largely on the temperature, and their fluorescence lifetime is weak. Fluorescence polarization is high, even at ambient temperature. At a low pH, the emission maximum is temperature-dependent.

   Puric bases fluoresce at least three times more than pyrimidic bases. Guanidine is the most fluorescent basis. In the native state, DNA and RNA show a very weak fluorescence,

Table 7.3 Absorption and spectral properties of some used fluorophores

Fluorophore	Binding site	Absorption		Emission		
		λ_{max} (nm)	ε_{max} ($\times 10^{-3}$)	λ_{max}	Φ_f	τ_f (ns)
Dansyl chloride	Covalent bond on Cys and Lys	330	3.4	510	0.1	13
1,5-I-AEDANS	Covalent bond on Lys and Cys	360	6.8	480	0.5	15
Fluorescein isothiocyanate (FITC)	Covalent bond on lysines	495	42	516	0.3	4
8-Anilino-1-naphthalene sulfonate (ANS)	Noncovalent bond on proteins	374	6.8	454	0.98	16
Pyrene and derivatives	Membranes	342	40	383	0.25	100
Ethidium bromide	Noncovalent bond on nucleic acids	515	3.8	600	1	26.5

Reproduced from Cantor, R.C. and Schimmel, P.R. (1980). *Biophysical Chemistry*, W.H. Freeman, New York. Fluorescence lifetimes are means.

to the difference of the yeast transfer RNA (t-RNA) that contains a highly fluorescent basis called Y basis.

At 77 K, the DNA and RNA fluorescence increases. This fluorescence is due to the formation of adenine and thymine dimers, while the fluorescence of guanine and cytosine is inhibited. Ethidium bromide, acridine and Hoechst 33258 intercalate between DNA and RNA bases, thereby inducing a fluorescence increase.

Ethidium bromide emits red fluorescence when bound to DNA. The fluorescent probe is trapped in the base pair of DNA (Figure 7.16), thereby inducing an increase in its fluorescence intensity. Also, ethidium bromide can bind to double-stranded and single-stranded DNA and to RNA.

Hoechst 33258 binds to the minor groove of double-stranded DNA with a preference for the A–T sequence (Pjura *et al.* 1987). Interaction between DNA and proteins very often induces structural modifications in both interacting molecules. Such modifications in DNA can be characterized with 2-aminopurine (2AP), which is a highly fluorescent isomer of adenine. 2AP does not alter the DNA structure. It forms a base pair with thymine and can be selectively excited, since its absorption is red-shifted compared to that of nucleic acids and aromatic amino acids. In addition, its fluorescence is sensitive to the conformational change that occurs within the DNA (Rachofsky *et al.* 2001).

7.3.3.4 Ions detectors

Some fluorophores are used to detect the presence of sodium or potassium in cells. For example, the maxima of the fluorescence excitation and emission spectra of SBFI increase, and its emission maximum shifts toward short wavelengths in the presence of sodium (Figure 7.17).

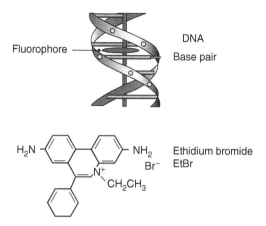

Figure 7.16 Binding of ethidium bromide to DNA.

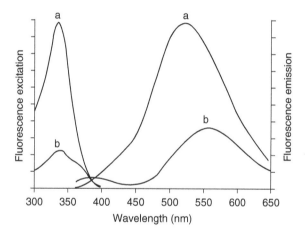

Figure 7.17 Fluorescence excitation (detected at 505 nm) and emission (excited at 340 nm) spectra of SBFI in pH 7.0 buffer containing 135 mM (a) or zero (b) Na^+. Source: Molecular Probes.

7.4 Polarity and Viscosity Effect on Quantum Yield and Emission Maximum Position

Quantum yield and fluorescence emission maximum are sensitive to the surrounding environment. This can be explained as follows. Fluorophore molecules and amino acids of the binding sites (in the case of an extrinsic fluorophore such as TNS, fluorescein, etc.) or the amino acids of their microenvironment (case of Trp residues) are associated by their dipoles. Upon excitation, only the fluorophore absorbs the energy. Thus, the dipole of the excited fluorophore has an orientation different from that of the fluorophore in the ground state. Therefore, the fluorophore dipole–solvent dipole interaction in the ground state is different from that in the excited state (Figure 7.18).

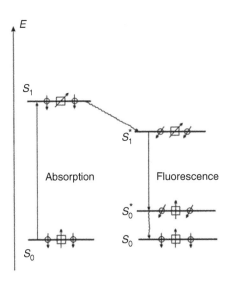

Figure 7.18 Relationship between fluorescence emission and dipole orientation. Squares characterize fluorophore dipole and circles solvent dipole.

The new interaction is unstable. To reach stability, fluorophore molecules will use some of their energy to reorient the dipole of the microenvironment (solvent and surrounding amino acids). Dipole reorientation is called relaxation. Emission occurs after the relaxation phenomenon.

An important dipole means a high polarity. In this case, the energy released by the fluorophore to reorient the dipole of the microenvironment is very important. Therefore, for a highly polar environment, photon emission will occur with low energy and thus with a spectrum maximum shifted to higher wavelength.

In an apolar medium, fluorophore in the excited state will induce a dipole formation within the environment. The formation of a new dipole needs less energy than the reorientation of an already existing dipole. Thus, emission from an apolar environment yields an emission spectrum with a maximum located in the blue compared to emission occurring in a polar environment.

Also, since the intensity and thus quantum yield are dependent on the number of emitted photons, they will be lower when emission occurs in a polar environment. It should be noted here that fluorescence parameters are more sensitive to the environment than absorption spectra.

When a fluorophore is bound to a protein, its fluorescence will be dependent on the polarity of the surrounding amino acids. Fluorescence spectra are also dependent on the rigidity of the medium. The relaxation phenomenon (reorientation of the dipole environment) occurs much more easily in a fluid medium. In such a case, emission will occur after relaxation. This is the case when relaxation is faster than fluorescence, i.e., the relaxation lifetime τ_r is shorter than the fluorescence lifetime τ_0. This occurs when the binding site is flexible, and the fluorophore can move easily. Emission from a relaxed state does not change with excitation wavelength. This can be explained by the fact that whatever the value of the excitation wavelength, the emission will always occur at the same energy level.

When the binding site is rigid, fluorescence emission occurs before relaxation. In this case, excitation at the longer wavelength edge of the absorption band photoselects the population of fluorophores energetically different from that photoselected when the excitation wavelength is shorter. In the absence of motions, specific excitation wavelength will reach specific population. Therefore, excitation at the red edge yields a fluorescence spectrum with a maximum located at a higher wavelength than that obtained when the excitation occurs at short wavelengths. In other words, when surrounding environment of the fluorophore is rigid, its emission maximum shifts to higher wavelengths with the excitation wavelength.

In summary, the shift of the emission peak could be the result of two phenomena: the decrease in the medium polarity and/or emission from a nonrelaxed state. Demchenko (1994) explains that:

> applying the term hydrophobic to fluorophores is not correct, fluorophores are rather amphipylic and posess a polar or even a charged group. Binding sites of these probes are not necessary hydrophobic; however numerous data demonstrate that their short wavelength shift of fluorescence spectra on binding is due to [the] slow rate of relaxations of the environment rather than to its hydrophobic nature.

References

Albani, J.R. (2004). *Structure and Dynamics of Macromolecules: Absorption and Fluorescence Studies*, Elsevier, Amsterdam.

Albani, J.R. (2007). New insights in the interpretation of tryptophan fluorescence. Origin of the fluorescence lifetime and characterization of a new fluorescence parameter in proteins: the emission to excitation ratio. *J. of fluorescence, in press.*

Badea, M.G. and Brand, L. (1979). Time-resolved fluorescence measurements. *Methods in Enzymology*, **61**, 378–425.

Balabanli, B., Kamisaki, Y., Martin, E. and Murad, F. (1999). Requirements for heme and thiols for the nonenzymatic modification of nitrotyrosine. *Proceedings of the Natural Academy of Sciences USA*, **96**, 13136–13141.

Barik, A., Priyadarsini, K.I. and Mohan, H. (2003). Photophysical studies on binding of curcumin to bovine serum albumin. *Photochemistry and Photobiology*, **77**, 597–603.

Bennett, R.G. (1960). Instrument to measure fluorescence lifetimes in the millimicrosecond region. *Review of Scientific Instruments*, **31**, 1275–1279.

Burstein, E.A., Vedenkina, N.S. and Ivkova, M. N. (1973). Fluorescence and the location of tryptophan residues in protein molecules. *Photochemistry and Photobiology*, **18**, 263–279.

Cantor, R.C. and Schimmel, P.R. (1980). *Biophysical Chemistry*, W.H. Freeman, New York.

Demchenko, A.P. (1994). A new generation of fluorescence probes exhibiting charge-transfer reactions. In: *Proceedings of SPIE – The International Society for Optical Engineering*, vol. 2137, 588–599.

Froschle, M., Ulmer, W. and Jany, K.D. (1984). Tyrosine modification of glucose dehydrogenase from *Bacillus megaterium*. Effect of tetranitromethane on the enzyme in the tetrameric and monomeric state. *European Journal of Biochemistry*, **142**, 533–540.

Handbook, Molecular Probes. 10th edn. (2005). http://probes.invitrogen.com/

Jablonski, A. (1935). Über den Mechanismus des Photolumineszenz von Farbstoffphosphoren. *Zeitschrift fur Physik*, **94**, 38–64.

James, D.R., Siemiarczuk, A. and Ware, W.R. (1992). Stroboscopic optical boxcar technique for the determination of fluorescence lifetimes. *Review of Scientific Instruments*, **63**, 1710–1716.

Lakowicz, J.R. (1999). *Principles of Fluorescence Spectroscopy* (2nd edn), Plenum, New York.

Pjura, P.E., Grzeskowiak, K. and Dickerson, R.E. (1987). Binding of Hoechst 33258 to the minor groove of B-DNA. *Journal of Molecular Biology*, **197**, 257–271.

Rachofsky, E.L., Osman, R. and Ross, J.B. (2001). Probing structure and dynamics of DNA with 2-aminopurine: effects of local environment on fluorescence. *Biochemistry*, **40**, 946–956.

Riordan, J.F., Wacker, W.E.C. and Vallee, B.L. (1966). Tetranitromethane. A reagent for the nitration of tyrosine and tyrosyl residues of proteins. *Journal of The American Chemical Society*, **88**, 4104–4105.

Schauenstein, E., Saenger, W., Schaur, R.J., Desoye, G. and Schreibmayer, W. (1980). Influence of pH, temperature and polarity of the solvent on the absorption of NADH+ and NADH at 260 nm. *Zeitschrift für Naturforschung C*, **35**, 76–79.

Spencer, R.D. (1970). Fluorescence lifetimes: theory, instrumentation and application of nanosecond fluorometry. Ph.D. thesis, University of Illinois at Urbana Champaign. Published by University Microfilms International.

Valeur, B. (2002). *Molecular Fluorescence: Principles and Applications*, Wiley-VCH, Weinheim.

Villette, J.R., Helbecque, N., Albani, J.R., Sicard, P.J. and Bouquelet, S.J. (1993). Cyclomaltodextrin glucanotransferase from *Bacillus circulans* E 192: nitration with tetranitromethane. *Biotechnology and Applied Biochemistry*, **17**, 205–216.

Chapter 8

Effect of the Structure and the Environment of a Fluorophore on Its Absorption and Fluorescence Spectra

Experiments

Students should read carefully the whole text before beginning the experiments and plotting the spectra. The purpose of these experiments is to find out how the fluorophore structure affects optical spectroscopy properties such as absorbance and fluorescence.

The following experiments have been chosen to provide students with experience with spectra of typical fluorescent molecules, which are often used in biochemical research. In particular, students should become familiar with the ways in which a molecule's environment can affect its energy states, and with the relation that exists between changes in energy states and the resulting modifications in absorption and fluorescence spectra.

Before entering the laboratory, students should be able to describe the following:

1. Absorption spectrum.
2. Emission spectrum.
3. Excitation spectrum.
4. Relationship between 1, 2, and 3.
5. Structures of ANS (1-anilinonaphthalene-8-sulfonate), L-tryptophan, riboflavin or flavin, ethidium bromide (EB), TNS, and NADH.

Students should receive the following solutions:

- L-Trp in pH 7 buffer with an optical density at 295 nm equal to 0.12.
- Riboflavin in pH 7 buffer with an optical density at 450 nm equal to 0.015.
- Ethidium bromide in pH 7 buffer with an optical density at 480 equal to 0.153.
- ANS in pH 7 buffer with an optical density at 350 nm equal to 0.06.
- ANS in the presence of 10 μM of bovin serum albumin (optical density at 350 nm should be around to 0.065).
- TNS in pH 7 buffer with an optical density at 320 nm equal to 0.06.
- TNS in the presence of 10 μM of bovin serum albumin (optical density at 320 nm should be around to 0.065).
- NADH dissolved in phosphate buffer pH 7 and in butylglycol (optical density should be around 0.04 at 280 nm).

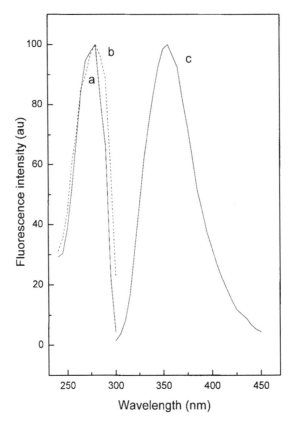

Figure 8.1   Normalized fluorescence emission spectrum (λ_{ex} = 295 nm with emission peak = 355 nm) (c), excitation spectrum obtained at λ_{em} = 340 nm (b), and absorption spectrum (a) of L-tryptophan in phosphate buffer, pH 7.

Plot on the same graph paper the absorption spectrum of L-Trp solution from 240 to 350 nm, the fluorescence emission spectrum from 300 to 450 nm (λ_{ex} = 295 nm), and the fluorescence excitation spectrum from 240 to 350 nm (λ_{em} = 340 nm).

Plot on the same graph paper the absorption spectrum of the riboflavin solution from 300 to 500 nm, fluorescence emission spectrum from 480 to 600 nm (λ_{ex} = 450 nm) and the fluorescence excitation spectrum from 300 to 500 nm (λ_{em} = 530 nm).

Plot on the same graph paper the absorption spectrum of ethidium bromide from 420 to 580 nm, its fluorescence emission spectrum from 500 to 700 nm (λ_{ex} = 475 nm) and its fluorescence excitation spectrum from 420 to 580 nm (λ_{em} = 600 nm).

Plot on the same graph paper the absorption spectra from 240 to 450 nm, fluorescence emission spectra from 360 to 550 nm (λ_{ex} = 350 nm), and fluorescence excitation spectra from 240 to 450 nm (λ_{em} = 480 nm) of ANS and of ANS–BSA solutions.

Plot on the same graph paper, the normalized excitation (from 290 to 420 nm) (λ_{em} = 440 or 450 nm) and emission spectra (from 360 to 560 nm) (λ_{ex} = 340 nm) of NADH dissolved

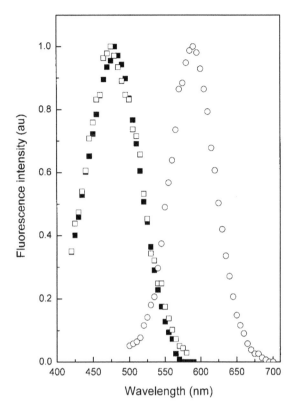

Figure 8.2 Normalized fluorescence emission spectrum (λ_{ex} = 475 nm) with emission peak at 585 nm (○), excitation spectrum (λ_{em} = 600 nm (□)) and absorption spectrum (■) of ethidium bromide in phosphate buffer, pH 7.

in phosphate buffer pH 7 and in butylglycol and compare for both solvents, excitation peaks together and emission peaks together.

For the solutions containing Trp, riboflavin, ethidium bromide, and ANS alone, normalize the absorption and excitation spectra and superimpose them. We suggest that the normalization be carried out at 280, 450, and 280 nm, for L-Trp, riboflavin, and ANS, respectively.

For the solution containing the ANS–BSA complex, normalize the absorption spectrum of free ANS in solution with the excitation spectrum of the ANS–BSA complex at 280 nm and superimpose them.

Questions:

1 Are the absorption spectra of the fluorophores identical? Explain.
2 Are the fluorescence emission spectra of these fluorophores identical? Explain.
3 Are the fluorescence excitation and absorption spectra of each fluorophore superimposed? Explain.

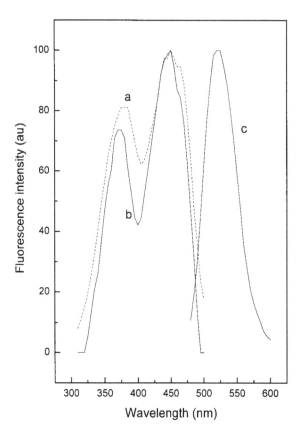

Figure 8.3 Normalized fluorescence emission spectrum (λ_{ex} = 450 nm with emission peak = 520 nm) (c), excitation spectrum (λ_{em} = 530 nm) (a) and absorption spectrum (b) of riboflavin in phosphate buffer, pH 7.

4 Are the fluorescence excitation spectrum of the ANS–BSA complex and the absorption spectrum of the free ANS in solution superimposed? Explain.

5 Are the fluorescence excitation and absorption spectra of the ANS–BSA complex superimposed? Explain.

6 What will happen if you correct the fluorescence intensities of emission and excitation spectra for the inner filter effect?

7 Calculate the Stokes shift for each fluorophore.

8 Plot the fluorescence emission spectra of ANS bound to serum albumin and of TNS (2, p-toluidinylnaphthalene-6-sulfonate) also bound to the protein. Are the two spectra identical? Explain.

9 Plot the fluorescence emission spectra of free TNS in solution (around 10–20 μM) in the absence and presence of the same concentration of free tryptophan. What do you observe? Can you explain the difference observed in the emission spectra obtained in parts 8 and 9?

10 Are the NADH spectra in good agreement with the polarity of the solvents used?

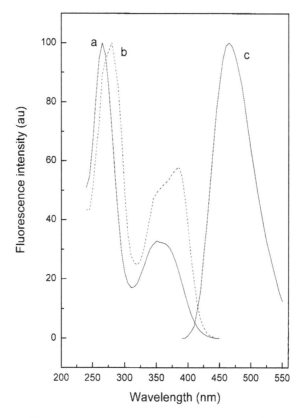

Figure 8.4 Normalized fluorescence emission spectrum (λ_{ex} = 350 nm with emission peak = 465 nm) (c), excitation spectrum (λ_{em} = 480 nm) (b) of ANS bound to BSA and absorption spectra (a) of free ANS in phosphate buffer, pH 7.

Answers:

1 The fluorophores show different absorption spectra since they possess different structures. In fact, the absorption spectrum reflects the electronic distribution of the molecule in the ground state and thus the molecule structure in the ground state.

2 The fluorophores show different fluorescence spectra. A fluorescence spectrum reflects the electronic distribution of the molecule in the excited state S_1 and thus the structure of the molecule in this state. We notice also that ANS does not fluoresce in a phosphate buffer (i.e., a polar medium), which is not the case when it is bound to serum albumin. Thus, fluorescence depends not only on the structure of the fluorophore but also on its environment. Each fluorophore has its own spectral properties, and one should be careful not to generalize a rule for all fluorophores, although most of the fluorophores share common rules.

3 Since ANS dissolved in a polar medium does not fluoresce, one cannot record its fluorescence excitation spectrum. For tryptophan, ethidium bromide, and riboflavin, one can see that for each molecule, the absorption spectrum looks like the fluorescence

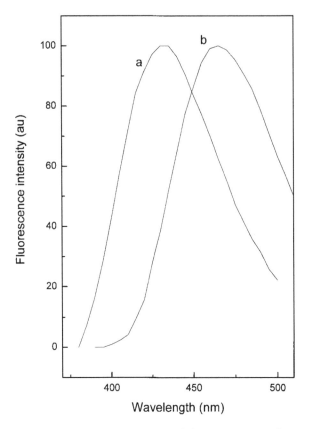

Figure 8.5 Normalized fluorescence emission spectra of the TNS–BSA complex ($\lambda_{ex} = 320$ nm and $\lambda_{em} = 432$ nm) (a) and of ANS–BSA complex ($\lambda_{ex} = 350$ nm and $\lambda_{em} = 465$ nm) (b).

excitation spectrum (Figures 8.1–8.3). In general, excitation and absorption spectra are supposed to superimpose when the fluorophore is in pure solution, and no dynamical or structural phenomena interface with it. This is true for fluorophores such as ethidium bromide but not for L-Trp and riboflavin. In fact, although the general feature of excitation spectrum looks like the absorption spectrum, there is an important difference between the spectra recorded for L-Trp and riboflavin: the intensity of the excitation spectrum does not vary with the wavelength in the same way them the optical density along the absorption spectrum. Detailed studies of the phenomenon have shown that absorption is not necessarily equal to excitation and thus fluorescence emission spectra of fluorophores such as aromatic amino are proportional to the intensity of the excitation spectrum at the excitation wavelength and not to the optical density at the excitation wavelength (Albani, 2007).

4   The absorption spectrum of free ANS in solution fluorescence excitation spectrum of the ANS–BSA complex do not overlap as a result of molecular interaction between ANS and BSA (Figure 8.4). Energy transfer occurs between Trp residues of the BSA and ANS molecules bound to the protein affecting the fluorescence excitation spectrum

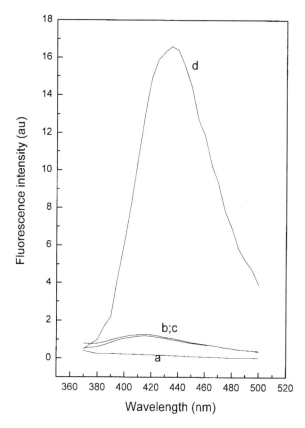

Figure 8.6 Emission spectra of buffer (a), free TNS (b), free TNS in the presence of free L-tryptophan (c) and of TNS bound to bovine serum albumin (d). $\lambda_{ex} = 320$ nm.

of bound ANS. In fact, one can see that in the excitation spectrum of the complex, there is a peak at 280 nm characteristic of the protein. One can see also that there is a shift in the second peak of ANS absorption from 350 to 385 nm in the excitation spectrum. This shift also characterizes the difference in the structural environment of ANS (buffer and protein).

5   The absorption spectrum of ANS bound to serum albumin shows a peak at 370 nm, different from 350 nm observed for ANS alone and 385 nm observed in the excitation spectrum. In general, the excitation spectrum is much more accurate than the absorption spectrum because it characterizes the real emitting fluorophores (here ANS bound to BSA). The absorption spectrum characterizes the sum of all absorbing molecules in the different ground states, here free and bound ANS.

6   Corrections of the fluorescence excitation and emission spectra for the optical densities (inner filter effect) yield an increase in the intensities without inducing any shift in any of the maxima. This is true for all recorded spectra. At the optical densities we are suggesting, you may find that these corrections are not significant and even not necessary for all fluorophores but ethidium bromide. Its intensity increases without any modification in the peak position.

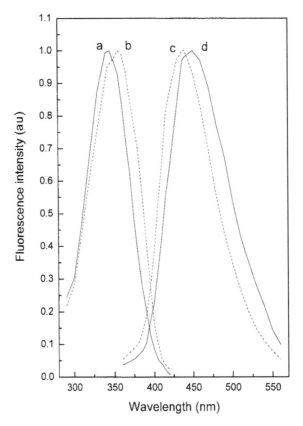

Figure 8.7 Fluorescence excitation (a and b) and emission (c and d) spectra of NADH dissolved in phosphate buffer, pH 7 (full lines), and in butylglycol (dashed lines). $\lambda_{ex} = 360$ nm and $\lambda_{em} = 440$ or 450 nm. Peaks positions are: in phosphate buffer, 345 and 450 nm and in butylglycol, 355 and 437 nm.

7   The Stokes shift is equal to the distance (nm) that exists between the absorption and emission peaks:

$$\text{Stokes shift} = \delta\lambda = \lambda_{em} - \lambda_{ex} \tag{8.1}$$

The Stokes shift is dependent on the structure of the fluorophore itself and on its environment (the solvent in which it is dissolved and/or the molecule to which it is bound. The Stokes shifts obtained are thus 75 nm (355–280 nm), 70 nm (520–450 nm), 113 nm (585–472 nm), and 185 nm (465–280 nm) for tryptophan, riboflavin, ethidium bromide, and ANS bound to BSA, respectively.

8   Figure 8.5 shows fluorescence emission spectra of TNS and ANS bound to serum albumin. The two fluorophores do not show the same maximum, although the two fluorophores bind to hydrophobic domains of the protein. This result can be explained mainly by the differences in the structure of the two fluorophores. Also, one could explain the difference in the emission peaks could be due to the higher sensitivity of TNS to hydrophobicity. This interpretation is based on the fact that the peak of TNS emission spectrum is shifted to short wavelengths compared to that of ANS

emission spectrum. However, this interpretation should be considered with caution, since effective comparative studies between the two fluorophores have not yet been undertaken. The fluorescence excitation spectrum of the TNS–BSA complex (not shown), recorded at λ_{em} = 460 nm, a wavelength where only TNS emits, shows a peak located at 278 nm, characteristic of the protein absorption. This peak, absent in the absorption spectrum of TNS dissolved in buffer, indicates, as in the ANS–BSA complex, energy-transfer between tryptophan residues of serum albumin and TNS. The fluorescence intensity of fluorophores such as TNS and ANS increases when they bind to a protein or to a membrane. In a polar solvent such as water, the two fluorophores mainly TNS, do not show any significant fluorescence.

9 Figure 8.6 clearly indicates that in the presence of free tryptophan in solution, there is no binding of TNS on the amino acid. However, in the presence of bovine serum albumin, TNS shows a fluorescence emission spectrum, indicating that TNS is bound to the protein.

10 Regarding fluorescence excitation and emission spectra of NADH, Figure 8.7 shows the normalized spectra obtained in phosphate buffer, pH 7, and in butylglycol. The excitation peak of NADH in phosphate buffer is located at 343–344 nm, while in butylglycol it is located at 355 nm. This shift to higher wavelengths of the excitation peak is the result of a low polarity of butylglycol compared to phosphate buffer. The difference in polarity also affects the emission position peak of NADH. In fact, emission peaks are located at 450 and 437 nm in phosphate buffer and butylglycol, respectively.

Reference

Albani, J.R. (2007). New insights in the interpretation of tryptophan fluorescence. Origin of the fluorescence lifetime and characterization of a new fluorescence parameter in proteins: the emission to excitation ratio. *J. of fluorescence, in press.*

Chapter 9

Fluorophore Characterization and Importance in Biology

This chapter contains four experiments, which are to be conducted separately.

9.1 Experiment 1. Quantitative Determination of Tryptophan in Proteins in 6 M Guanidine

9.1.1 Introduction

The luminescent properties of tyrosine and tryptophan are affected by their microenvironments. Depending upon the location within the polypeptide structure and the neighboring charges, several classes of residues have been distinguished in native proteins (Cogwill 1968; Burstein *et al.* 1973). Each class exhibits mainly a distinct spectral emission maximum with a specific spectral bandwidth. Nevertheless, upon denaturation, the protein unfolds, and, as was first reported by Teale (1960), for most proteins, the quantum yields of tryptophan residues tend to become uniform.

Fluorescence spectra of the tryptophanyl residues of the fully denatured protein is normalized, as is its absorption spectra. In the absence of any structured protein, fluorescence of tryptophanyl residues and of free Trp in solution can be compared.

With an efficient denaturing agent such as guanidine hydrochloride (Tanford *et al.* 1966, 1967) and in the presence of 2-mercaptoethanol, a more uniform environment could be reached and hence a normalization of the fluorescence parameters, a condition which has to be met before applying the process for quantitative analytical purposes.

9.1.2 Principle

The analytical method proposed by Pajot (1976) involves (a) incubating the protein of a known concentration in 6 M guanidine at pH 6.5–7 in the presence of 30 mM 2-mercaptoethanol for 30 min (if only a small amount of the protein is available, the incubation can be carried out directly in the fluorescence cuvette); and (b) measuring the fluorescence, excited at 295 nm and observed at 354 nm, yielded by the denatured protein in 6 M guanidine hydrochloride (tryptophanyl residues concentration 3–10 μM). A linear calibration curve is obtained, thus allowing the 'free tryptophan fluorescence equivalents' of the protein sample to be estimated by extrapolation.

Therefore, by carrying out a titration as described above, the fluorescence of the denatured protein can be compared to that of a given amount of free tryptophan. Plotting the fluorescence intensity at the emission maximum as a function of added free tryptophan residues allows the tryptophan concentration in the protein to be determined.

The fluorescence estimation may be impaired by the inner filter effect, but this may be made practically negligible by lowering the sample concentration. A maximal total concentration of 25 μM tryptophan is correct, since about 98% of the incident light is then available at the center of a 4-mm cell.

Although denaturation and the highly charged medium tend to make the fluorophore environment uniform, certain features of the primary structure may exert some quenching or enhancing effects on the fluorescence parameters of certain residues. This may be the case with lysozyme, which contains a Trp–Trp segment, or with myoglobin, in which a carboxylic charge is located on the amino acid adjacent to tryptophan. Such factors may be responsible for the deviation observed with some proteins. This type of study also relies upon the accuracy of the absorbance coefficients used for estimating protein concentrations.

9.1.3 Experiment

Prepare a 100 μM stock solution of L-Trp dissolved in 10 mM phosphate buffer. $\varepsilon_{279\,nm}$ of L-tryptophan is $5710\,\mathrm{M}^{-1}\,\mathrm{cm}^{-1}$. Prepare stocks of 10 different proteins in 10 mM phosphate buffer pH 7.

The concentration of each protein stock should be 25 μM. Incubate for 30 min, by slowly shaking 40 μl of each protein stock in a fluorescence cuvette containing 860 μl of 6 M guanidine, pH 7. The protein concentration in the cuvette is around 1 μM.

The stock of guanidine solution should be prepared in 0.1 M phosphate buffer.

After 30 min of incubation, record the fluorescence spectrum for each protein from 310 to 440 nm, $\lambda_{ex} = 295$ nm. Then, add to the cuvette at least 5 aliquots of 5 μl each from the L-Trp stock solution, and after each addition, record the fluorescence spectrum.

Plot the fluorescence intensity at the peak as a function of added L-Trp concentration, and then determine the concentration of the tryptophan in the protein. From the concentration of the protein you are using, you can determine the number of tryptophan present in each protein.

We suggest performing the experiments on the following proteins: rabbit muscle aldolase, $A_{280}^{1\%} = 9.38, M_r = 142\,000$; pig-heart aspartate aminotransferase, $\varepsilon_{280\,nm} = 1.40 \times 10^5\,\mathrm{M}^{-1}\,\mathrm{cm}^{-1}$ (dimer) and its apoenzyme $\varepsilon_{280\,nm} = 1.33 \times 10^5\,\mathrm{M}^{-1}\,\mathrm{cm}^{-1}$; bovine serum albumin, $\varepsilon_{280\,nm} = 4.2 \times 10^4\,\mathrm{M}^{-1}\,\mathrm{cm}^{-1}$; bovine α-chymotrypsin, $A_{282}^{1\%} = 20.4$, $M_r = 25\,600$; bovine α-chymotrypsinogen, $A_{280}^{1\%} = 21, M_r = 25\,600$; horse-heart cytochrome *c*, type VI, $\varepsilon_{550\,nm} = 29.5 \times 10^3\,\mathrm{M}^{-1}\,\mathrm{cm}^{-1}$; porcine cytochrome b_5, $\varepsilon_{556\,nm} = 27.5 \times 10^3\,\mathrm{M}^{-1}\,\mathrm{cm}^{-1}$; bovine deoxyribonuclease, $A_{280}^{1\%} = 12.3, M_r = 31\,000$; bovine hemoglobin ($Fe^{2+}$), $\varepsilon_{555\,nm} = 12.5 \times 10^3\,\mathrm{M}^{-1}\,\mathrm{cm}^{-1}$; muscle lactate dehydrogenase, $\varepsilon_{280\,nm} = 20.1 \times 10^3\,\mathrm{M}^{-1}\,\mathrm{cm}^{-1}$; β-lactoglobulin, $\varepsilon_{278\,nm} = 17.6 \times 10^3\,\mathrm{M}^{-1}\,\mathrm{cm}^{-1}$; egg lysozyme, $\varepsilon_{280\,nm} = 37.8 \times 10^3\,\mathrm{M}^{-1}\,\mathrm{cm}^{-1}$; sperm-whale myoglobin, $\varepsilon_{409\,nm} = 157 \times 10^3\,\mathrm{M}^{-1}\,\mathrm{cm}^{-1}$; porcine pepsin, $\varepsilon_{278\,nm} = 50.9 \times 10^3\,\mathrm{M}^{-1}\,\mathrm{cm}^{-1}$; bovine ribonuclease,

$\varepsilon_{277.5\,nm} = 9800 \text{ M}^{-1} \text{ cm}^{-1}$; trypsin, $A_{280}^{1\%} = 15.4$, $M_r = 23\,891$; soybean trypsin inhibitor, $A_{280}^{1\%} = 9.1$; $M_r = 21\,700$.

9.1.4 Results obtained with cytochrome b₂ core

Figure 9.1 shows the fluorescence spectrum of tryptophan residues of cytochrome b_2 core dissolved in 6 M guanidine, pH 7, in the presence of 30 mM 2-mercaptoethanol. The emission peak is located at 357 nm, indicating that the protein is completely denatured (spectrum a). The fluorescence emission spectra of successive aliquots (0.55 μM) of free tryptophan added to the protein solution are also shown (spectra b–g).

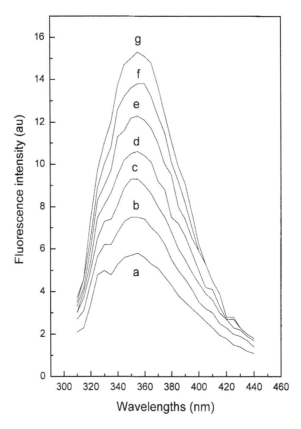

Figure 9.1 Titration of 1 μM cytochrome b_2 core in 6 M guanidine with aliquots of 5.5 μM of L-Trp. $\lambda_{ex} = 295$ nm.

Figure 9.2 shows the fluorescence intensity maximum as a function of added free tryptophan. Extrapolation of the plot at the x-axis yields a value equal to 2 μM as the concentration of tryptophan in the cytochrome b_2 core. Since the protein concentration in the cuvette is 1 μM, the number of tryptophan residues in the cytochrome b_2 core is

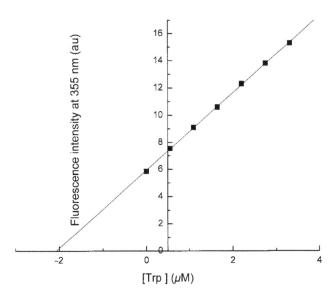

Figure 9.2 Determination of tryptophan residues concentration in the cytochrome b$_2$ core.

$2\ \mu M/1\ \mu M = 2$. This value is identical to that known for cytochrome b$_2$ core. In general, one should divide the value obtained by a correlation coefficient equal to 1.1 to obtain the real number of tryptophan residues in the protein.

9.2 Experiment 2. Effect of the Inner Filter Effect on Fluorescence Data

9.2.1 Objective of the experiment

Student will apply the Beer–Lambert–Bouguer law to find out the effect of the optical density on the fluorescence intensity. This experiment can be performed with any fluorophore, but students should be sure that it is not aggregating beyond a specific concentration. This is why it is better to perform this experiment with a fluorophore that has a high extinction coefficient, so that one can work at low concentrations. We have chosen fluorescein to perform the experiments.

9.2.2 Experiment

The same cuvette with each fluorescein concentration should be used to measure ODs at 495 nm and record the fluorescence intensities at 540 nm ($\lambda_{ex} = 495$ nm). Prepare a stock solution of at least 40 μM fluorescein. Calculate the volume you need to pipette and to add to the fluorescence cuvette containing the buffer alone (1 ml of buffer for example) so that you add aliquots of approximately 0.2 μM fluorescein to the cuvette.

Mix the solution, and then measure the OD and fluorescence intensity. Add aliquots until you have 4 μM fluorescein in the cuvette. Plot the OD at 495 nm vs. fluorescein concentration. Do you obtain a linear plot? What is the value of ε you obtain from your data?

Plot the fluorescence intensity at 540 nm vs. the OD at 495 nm. How can you interpret the plot you obtained? With Equation (7.15), calculate the value of the fluorescence intensity at each OD. Superimpose your theoretical results with the experimental plot. What do you observe?

9.2.3 Results

Figure 9.3 shows the variation in OD for fluorescein at 495 nm as a function of fluorophore concentration. We obtain a linear plot indicating that the Beer–Lambert–Bouguer law is respected in this range of fluorescein concentrations. Since, in the present measurements optical densities were measured with a pathlength equal to 1 cm, we obtained at 495 nm from the slope of the plot, a value equal to 42 mM^{-1} cm^{-1} for ε. This result, identical to

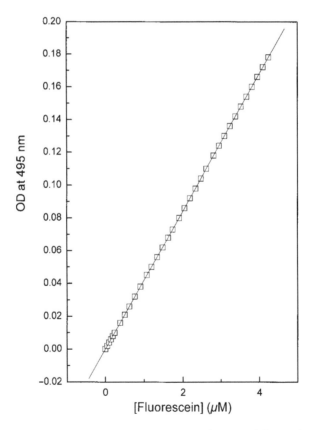

Figure 9.3 Optical density (OD) of fluorescein at 495 nm as a function of chromophore concentration.

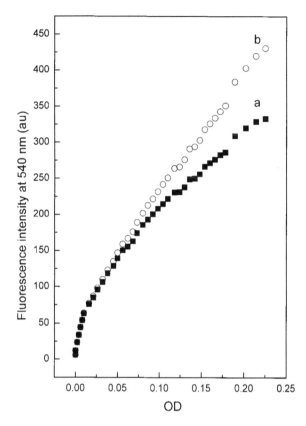

Figure 9.4 Fluorescence intensity of fluorescein at 540 nm as a function of the optical density at the excitation wavelength (495 nm). Plot (a) is the experimental, and plot (b) is calculated using Equation (7.15).

that known for fluorescein (see Table 7.3) means that the errors on the volume added and thus on the fluorescein concentrations in the cuvette are very small.

Plotting the fluorescein fluorescence intensity at 540 nm vs. the OD at the excitation wavelength (495 nm) does not yield a linear plot (Figure 9.4). Although we do not have an explanation for the small curvarture at low fluorescein concentrations, results indicate that linearity between fluorescence intensities and ODs are clearly lost when the OD is higher than 0.075. Corrections of the fluorescence intensities for the ODs at 495 nm (the inner filter effect) indicate that up to an OD of 0.075, the theoretical fluorescence intensities are equal to the experimental intensities. When ODs are equal to or higher than 0.075, theoretical fluorescence intensities are higher than the experimental intensities and are linear with the fluorescence intensities recorded for the lower ODs.

One can conclude that Equation (7.15) is very useful in correcting fluorescence intensities recorded at high ODs. Also, it is important to indicate that the curvature observed, at high ODs, in the experimental plot is not the result of fluorescein aggregation, since ODs increase linearly with fluorescein concentrations (Figure 9.3).

9.3   Experiment 3. Theoretical Spectral Resolution of Two Emitting Fluorophores Within a Mixture

9.3.1   Objective of the experiment

The objective of these experiments is to apply a method of calculation that allows, when possible, separation of the fluorescence spectra of two or three emitting residues. Instead of using a computer and applying specific programs, students will learn how to perform the mathematical calculation and will see that it is very simple to obtain the different spectra.

The method we are asking the students to use is the ln-normal analysis of Burstein and Emelyanenko (1996).

In this analysis, the following equation is used:

$$I_{(\nu)} = I_m \exp\{-(\ln 2/\ln^2 \rho) \times \ln^2[(a - \nu)/(a - \nu_m)]\} \tag{9.1}$$

where $I_{\nu_m} = I_m$ is the maximal fluorescence intensity. ν_m is the wavenumber of the band maximum,

$$\rho = (\nu_m - \nu_-)/(\nu_+ - \nu_m) \tag{9.2}$$

is the band asymmetry parameter. ν_+ and ν_- are the wavenumber positions of the left and right half maximal amplitudes:

$$a = \nu_m + H\rho/(\rho^2 - 1) \tag{9.3}$$

and H is the bandwidth:

$$H = \nu_+ - \nu_- \tag{9.4}$$

Equation (9.1) allows us in principle to draw a spectrum that matches the entire spectrum obtained experimentally. This is correct only if the recorded spectrum originates from one fluorophore (i.e., Trp in solution) or from compact protein within a folded structure. However, when the experimental spectrum originates from two fluorophores (i.e., mixtures of tyrosine and tryptophan in solution) or from a disrupted protein that has two classes of Trp residues, the calculated spectrum using Equation (9.1) does not match the recorded spectrum.

- Step 1: Students will record the fluorescence emission spectra of 10 μM L-Trp, 10 μM L-Tyr, and around 160 μM Calcofluor.
- Step 2: Students will record the fluorescence spectrum of a (L-Trp + L-Tyr) mixture.
- Step 3: Students will analyze emission spectra obtained in steps 1 and 2 using the Burstein equation. For the fluorescence spectra recorded in step 1, the Burstein equation should yield theoretical spectra that match the recorded spectra. However, this will not be the case for the tyrosine–tryptophan mixture.
- Step 4: Students will plot the fluorescence emission spectrum of the Trp residues of a protein.
- Step 5: Students will analyze the emission spectra obtained in step 4 with the Burstein equation. Correct the fluorescence spectrum for the inner filter effect before doing the analysis with the Burstein–Emelyanenko method.

We will show here the example of an analysis method concerning the fluorescence emission of Calcofluor. First, with a Pasteur pipette, take a small amount of powder of Calcofluor White and dissolve it in 1 ml of phosphate buffer present in the fluorescence cuvette. Then, measure the OD at 352.7 nm and calculate the fluorophore concentration in the cuvette. The value of ε at this wavelength is 4388 M^{-1} cm^{-1}. In principle, you should have a Calcofluor solution concentration of 150–200 μM in the fluorescence cuvette.

Plot the fluorescence emission spectrum from 380 to 550 nm (λ_{ex} = 300 nm). You will obtain the spectrum shown in Figure 9.5.

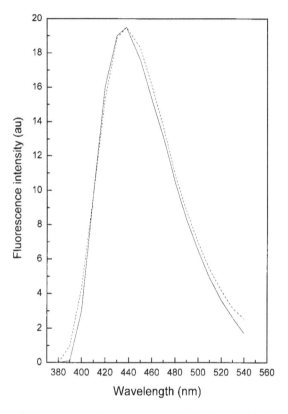

Figure 9.5　Experimental fluorescence emission spectrum of 160 μM calcofluor in water (full line) λ_{ex} = 300 nm, and that calculated theoretically (dashed line) using Equation (10.1).

Let us see the details of the equation we are going to use

$$I_{(v)} = I_m \exp\{-(\ln 2/\ln^2 \rho) \times \ln^2[(a - v)/(a - v_m)]\} \tag{9.1}$$

I_m is the maximal intensity of the spectrum, i.e., at 438 nm, and is 19.5.

$$v_m = 1/438 \text{ nm} = 22.83 \times 10^3 \text{ cm}^{-1}$$

$$v_- = 1/434 \text{ nm} = 20.66 \times 10^3 \text{ cm}^{-1}$$

$$v_+ = 1/410 \text{ nm} = 24.4 \times 10^3 \text{ cm}^{-1}$$

$H = v_+ - v_- = 3.8 \times 10^3 \text{ cm}^{-1}$

$\rho = (v_m - v_-)/(v_+ - v_m) = (22.83 - 20.66)/(24.4 - 22.83) = 2.17/1.57 = 1.382.$

$a = v_m + H\rho/(\rho^2 - 1) = 22.83 + (3.8 \times 1.382)/0.909924 = 28.6 \times 10^3 \text{ cm}^{-1}$

Now that we have all the parameters values we need, we can replace them in Equation (9.1) and determine the value of the intensity $I_{(v)}$ at a specific wavenumber (v).

$$I_{(400\,nm)} = 19.5 \exp\{-(0.693/0.1046) \times \ln^2[(28.6 - 25)/(28.6 - 22.83)]\}$$

$$= 19.5 \exp[-6.625 \times \ln^2(3.6/5.8)]$$

$$= 19.5 \times 0.2216 = 4.32 \text{ cm}$$

The intensity measured at 400 nm from the spectrum is 3 cm, and that calculated theoretically is 4.32 cm. One can see that the two values are close. Let us do this along the emission wavelength:

λ (nm)	I measured from the recorded spectrum (cm)	I calculated theoretically (cm)
400	3	4.32
410	9.8	9.77
420	16	15.3
430	19	18.86
438	19.5	19.5
450	17.6	18.3
460	15.3	16.1
470	13.1	13.7
480	10.5	10.9
490	8.3	8.8
500	6.5	6.97
510	4.9	5.4
520	3.6	4.17
530	2.6	3.15
540	1.7	2.54

Now, let us plot the two spectra on the same graph. We shall obtain two identical spectra with one peak at 438 nm (Figure 9.5).

9.3.2 Results

Figure 9.6 shows the emission spectrum of free L-Trp in solution.

Figure 9.7 shows the experimental fluorescence spectrum of a mixture of L-tyrosine and L-tryptophan in water (line, spectrum a). The presence of a small shoulder around 303 nm indicates that L-tyrosine contributes to the fluorescence emission. Analysis of the data using

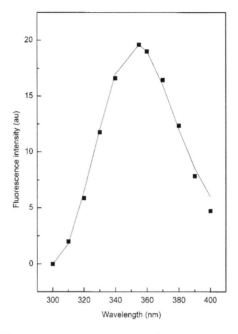

Figure 9.6 Experimental fluorescence emission spectrum of pure L-Trp in water (line) and that obtained using Equation (10.1) (■).

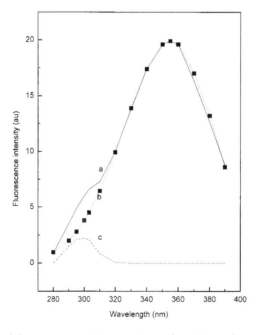

Figure 9.7 Experimental fluorescence emission spectrum of a mixture of L-tyrosine and L-tryptophan in water (line, spectrum a) and that approximated using Equation (9.1) (■, spectrum b). Substracting spectrum (b) from (a) yields a fluorescence spectrum (c) characteristic of tyrosine. $\lambda_{ex} = 260$ nm.

Equation (9.1) yields a fluorescence spectrum (squares, spectrum b) that does not match with the experimental one, especially within the region where tyrosine emits. The calculated spectrum (b) corresponds to that of tryptophan alone. Subtracting spectrum (b) from (a) yields a spectrum with a maximum at 303 nm corresponding to that of tyrosine.

Figure 9.8 shows the experimental fluorescence spectrum of Trp residues of 15 μM α_1-acid glycoprotein (λ_{ex} = 295 nm) (line) and calculated spectrum obtained using Equation (9.1) (square). We notice that the calculated spectrum matches the experimental spectrum. Note: students can apply this method to different fluorophore mixtures.

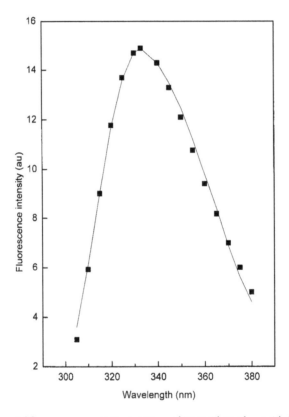

Figure 9.8   Experimental fluorescence emission spectrum of Trp residues of α_1-acid glycoprotein (15 μM) (line) and that approximated using Equation (10.1) (■). λ_{ex} = 295 nm.

9.4 Experiment 4. Determination of Melting Temperature of Triglycerides in Skimmed Milk Using Vitamin A Fluorescence

9.4.1 *Introduction*

Ultraviolet absorption spectrum of vitamin A shows a maximum near 328 nm, which was used extensively as a means for both identification and determination of the vitamin in the various natural sources in which it occurs. Different solvents such as alcohol, chloroform, benzene, and light petroleum have been used to study the effects of these solvents on the spectral properties of vitamin A from different oils and butter (Gillam and El Ridi 1938).

Vitamin A shows a fluorescence emission spectrum with a peak located at 450 nm. Since fluorescence is more sensitive than absorption, vitamin A is retained in many foods such as butter, oil, milk, and cheese, which retain the important quantities of vitamin A, and can be studied by fluorescence spectroscopy. However, because of the thickness and density of such products, classical right-angle spectroscopy studies are difficult to perform. In fact, high ODs at excitation and emission wavelengths may not only decrease the real fluorescence intensity of the sample but also distort fluorescence spectra. We observed this fact when we tried to record the fluorescence emission spectrum of pure milk (spectrum not shown). In order to circumvent this problem, one should use front-face fluorescence spectroscopy. In this method, the sample is excited at the cuvette surface, thereby avoiding any displacement of the excitation light through the sample to the cuvette center (Figure 9.9). Therefore, the front-face technique yields fluorescence spectra (excitation and emission) that are not distorted. In fact, with highly concentrated and opaque liquids, intensity measurements are not reproducible or detectable. In the front-face technique, the excitation light is focused to the front surface of the samples, and then fluorescence emission is collected from the same region at an angle that minimizes reflected and scattered light. Front-face spectroscopy is applied to highly concentrated opaque or solid samples.

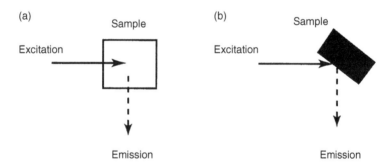

Figure 9.9　Principles of (a) right-angle and (b) front-face fluorescence spectroscopies.

The front-face technique was used by Dufour (2002) to study the melting point of a milk fat-in-water emulsion stabilized by β-lactoglobulin. Milk and milk products retain vitamin A located in the core and membrane of the fat globule. The lipids of milk fat globules contain hundreds of triacylglycerol species. Increasing the temperature modifies the textural structures of these fat molecules and thus affects the fluorescence of vitamin A. Thus, vitamin A can be used as probe to follow structural changes within the fat particles and to study, if possible, their melting point. Figure 9.10 shows the excitation spectra of a milk fat-in-water emulsion at different temperatures and the ratios of the intensities at 322 and 295 nm. The shape of the excitation spectra changes with temperature (Figure 9.10a). The modification of the ratio I_{322}/I_{295} with the temperature (Figure 9.10b), allows to determine the melting point of the triglycerides found to be equal to 28°C.

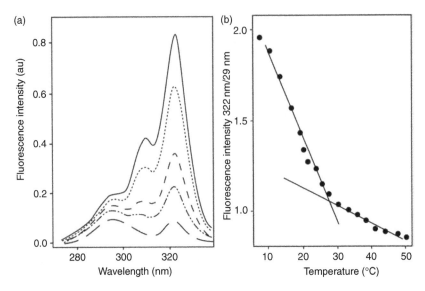

Figure 9.10   (a) Excitation spectra of vitamin A of a milk fat-in-water emulsion (melting point of the fats: 28°C) recorded at (—) 8.3°C, (···) 13.8°C, (---) 20.6°C, (----) 30.2°C, and (——) 49.1°C. (b) Changes in the ratio F322 nm/295 nm with temperature. Figure reprinted with permission from Dufour, E. (2002). *American Laboratory*, 2202 **34**, 51–55. Copyright © 2002 International Scientific Communnications, Inc.

9.4.2   Experiment to conduct

The experiment described in Figure 9.10 can be reproduced if front-face fluorescence is used. Otherwise, another experimental method should be applied to determine melting point of triglycerides. The following method can be used. Add 15–20 μl of skimmed milk into 1 ml of 1 N NaOH. The milk we used is enriched with vitamins E and B. By mixing slowly, you should soon obtain a clear and slightly pink solution.

Measure the fluorescence excitation intensity at 295 nm for increasing temperatures from 7 to 40°C, $\lambda_{em} = 410$ nm. Then, plot the fluorescence intensity vs. temperature. You should obtain an exponential decay curve. Finally, plot ln I vs. $t°$ and calculate the melting point.

9.4.3   Results

The fluorescence excitation spectrum of the milk–NaOH solution shows two maxima at 280 and 325 nm (data not shown). Figure 9.11 shows the variation in fluorescence intensity at $\lambda_{ex} = 295$ nm vs. temperature. The results clearly indicate that the fluorescence intensity is temperature-dependent and decreases exponentially with temperatures. Experiments should be repeated five times and performed on two different days. From the plot obtained, it is difficult to calculate the melting point of the triglycerides.

Therefore, we plot ln I vs. temperature. This should yield one straight slope (in the absence of a melting point) or two slopes separated by an area where the slope is almost equal to zero. This area characterizes the transformation of the triglycerides from a solid-like state to a liquid state. Figure 9.12 clearly shows that we are in the presence of a melting area, indicating that fusion of triglycerides is not abrupt.

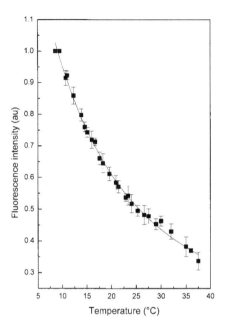

Figure 9.11 Fluorescence intensity decrease for 20 μl of skimmed milk ("Candia, Silhouette") containing 0.1 g of lipid in each 100 ml, dissolved in 1 ml of 1 N NaOH. $\lambda_{ex} = 295$ nm and $\lambda_{em} = 410$ nm. The data shown are mean values from five experiments.

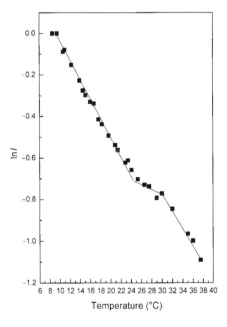

Figure 9.12 Fluorescence intensity decrease for 20 μl of skimmed milk ("Candia, Silhouette") dissolved in 1 ml of 1 N NaOH, expressed as ln I. $\lambda_{ex} = 295$ nm and $\lambda_{em} = 410$ nm. The data shown are mean values from five experiments. The plateau observed between 24 and 30 °C indicates the presence of a melting area with a mean melting temperature at 27 °C.

This result is in good agreement with that found by Dufour, and shows that alternative methods to front-face fluorescence could be used. Our results indicate that a melting area exists, and not only a single melting temperature. This means that structural changes occur within a temperature range where triglycerides are in a semi-solid–semi-fluid state. NaOH is used in general to neutralize the acid pH of milk. Finally, it is important to mention that we have not been able to dissolve non-skimmed (complete) milk in 1 N NaOH.

References

Burstein, E.A. and Emelyanenko, V.L. (1996). Log-normal description of fluorescence spectra of organic fluorophores. *Photochemistry and Photobiology*, **64**, 316–320.

Burstein, E.A., Vedenkina, N.S. and Ivkova, M.N. (1973). Fluorescence and the location of tryptophan residues in protein molecules. *Photochemistry and Photobiology*, **18**, 263–279.

Cogwill, R.W. (1968). Fluorescence and protein structure : XV. Tryptophan fluorescence in a helical muscle protein. *Biochimica et Biophysica Acta*, **168**, 432–438.

Dufour, E. (2002). Examination of the molecular structure of food products using front-face fluorescence spectroscopy. *American Laboratory*, **34**, 51–55.

Gillam, A.E. and El Ridi, M.S. (1938). The variation of the extinction coefficient of vitamin A with solvent. *The Biochemical Journal*, **32**, 820–825.

Pajot, P. (1976). Fluoescence of proteins in 6-M guanidine hydrochloride. A method for the quantitative determination of tryptophan. *European Journal of Biochemistry*, **63**, 263–269.

Tanford, C., Kawahara, K. and Lapanje, S. (1966). Proteins in 6 M Guanidine Hydrochloride. Demonstration of random behaviour. *Journal of Biological Chemistry*, **241**, 1921–1923.

Tanford, C., Kawahara, K. and Lapanje, S. (1967). Proteins as random coils. I. Intrinsic viscosities and sedimentation coefficients in concentrated guanidine hydrochloride. *Journal of the American Chemical Society*, **89**, 729–736.

Teale, F.W.J. (1960). The ultraviolet fluorescence of proteins in neutral solution. *The Biochemical Journal*, **76**, 381–388.

Chapter 10
Fluorescence Quenching

10.1 Introduction

Fluorescence is characterized by parameters such as intensity, quantum yield, and lifetime. The fluorescence intensity at a given wavelength is equal to the number of photons emitted by fluorescence multiplied by the photon energy:

$$I_F = nE \tag{10.1}$$

The fluorescence quantum yield Φ_F is equal to the number of photons emitted by fluorescence divided by the number of absorbed photons. In general, I_F and Φ_F are proportional to each other. The fluorescence lifetime τ_0 is the time spent by the fluorophore in the excited state.

External molecules added to the fluorescent system can quench fluorescence intensity and therefore quantum yield. These quenchers will decrease the fluorescence while entering in collision with the fluorophore (e.g., oxygen molecules that are going to diffuse in a protein and to enter in collision with the fluorescent Trp residues) or while forming a nonfluorescent complex with the fluorophore (e.g., a ligand that is going to bind on a protein, inducing quenching of the Trp residues fluorescence). In the first case, we speak of dynamic quenching, and in the second case of static quenching. Before we go further in the mathematical details of the two types of quenching, let us explain in simple words how they will affect the three fluorescence parameters mentioned above (intensity, lifetime, and quantum yield). Let us consider that we have 100 fluorophore molecules. At a specific emission wavelength, the intensity can be defined by Equation (10.1), i.e., the measured intensity is proportional to the number of emitted photons. Upon dynamic or static quenching, there will be fewer photons emitted, and so the emitted intensity will decrease. Since the quantum yield and intensity are proportional, a decrease in fluorescence intensity is accompanied by a decrease in quantum yield.

The fluorescence lifetime is the same for all 100 fluorophore molecules present in solution. When static quenching occurs, some of the fluorophore molecules bind to the ligand, and the formed complex does not fluoresce. Thus, upon static quenching, the number of molecules that are still fluorescing, i.e., not participating in the complex, is less than 100. However, all these molecules have the same fluorescence lifetime. Therefore, the mean fluorescence lifetime calculated in the presence of static quenching will be equal to that measured in the absence of the ligand. Thus, when static quenching occurs, the fluorescence intensity and quantum yield decrease, and the fluorescence lifetime does not change. Direct physical interaction between the fluorophore and the ligand is not necessary to observe static quenching. We can observe a long- or short-distance effect, depending on whether the fluorophore is within the interaction area or not.

Dynamic quenching does not induce complex formation; the fluorophore and quencher collide, inducing a loss of fluorophore energy; and then the two molecules are again separated. Thus, dynamic quenching induces a partial energy loss of the fluorophore molecules entering into collision with the quencher molecules. This means that the fluorescence lifetime of these molecules is lower than for those free in solution and not participating in the dynamic process. Therefore, the mean fluorescence lifetime measured in the presence of dynamic quenching will be lower than the mean lifetime measured in the absence of the ligand. In conclusion, dynamic quenching induces a decrease in fluorescence intensity, quantum yield, and lifetime. Also, one can observe a thermal quenching, which is the decrease in the fluorescence parameters (lifetime, intensity and quantum yield) with temperature. This decrease is dependent on the dynamics of the surrounding environment of the fluorophore.

Förster energy transfer or energy transfer at a distance occurs between two molecules, a donor (the excited fluorophore), and an acceptor (a chromophore or fluorophore). Energy is transferred by resonance, i.e., the electron of the excited molecule induces an oscillating electric field that excites the acceptor electrons. As a result of this energy transfer, the fluorescence intensity and quantum yield of the emitter will decrease. Energy transfer is described in Chapter 14.

10.2   Collisional Quenching: the Stern–Volmer Relation

Macromolecules display continuous motions. These motions can be of two main types: the molecule can rotate on itself, following the precise axis of rotation, and it can have a local flexibility. Local flexibility, also called internal motions, allows different small molecules, such as solvent molecules, to diffuse along the macromolecule. This diffusion is generally dependent on the importance of the local internal dynamics. Also, the fact that solvent molecules can reach the interior hydrophobic core of macromolecules such as proteins clearly means that the term hydrophobicity should be considered as relative and not as absolute. Internal dynamics of proteins allow and facilitate a permanent contact between protein core and the solvent. Also, this internal motion permits small molecules such as oxygen to diffuse within the protein core. Since oxygen is a collisional quencher, analyzing the fluorescence data in the presence of different oxygen concentrations yields information on the internal dynamics of macromolecules.

Dynamic quenching occurs within the fluorescence lifetime of the fluorophore, i.e., during the excited-state lifetime. This process is time-dependent. We have defined fluorescence lifetime as the time spent by the fluorophore in the excited state. Collisional quenching is a process that will depopulate the excited state in parallel to the other processes already described in the Jablonski diagram. Therefore, the excited-state fluorescence lifetime is lower in the presence of a collisional quencher than in its absence.

The velocity of fluorophore de-excitation can be expressed as

$$v = k_q[F][Q] \tag{10.2}$$

where k_q is the bimolecular quenching constant (M^{-1} s^{-1}), $[F]$ is the fluorophore concentration, which is held constant during the experiment, and $[Q]$ is the quencher concentration. If $[Q]$ is higher than $[F]$, the system can be considered as pseudo-first order with a constant equal to $k_q \times [Q]$.

$$v = k'[F] = k_q[Q][F] \tag{10.3}$$

The quantum yield in the absence of the quencher is

$$\Phi_F = \frac{k_r}{k_r + k_i + k_{isc}} \tag{10.4}$$

where k_r is the radiative constant, k_{isc} the intersystem crossing constant, and k_i the constant corresponding to de-excitation due to the temperature effect.

In the presence of a quencher, the quantum yield is

$$\Phi_{F(Q)} = \frac{k_r}{k_r + k_i + k_{isc} + k_q[Q]} \tag{10.5}$$

$$\frac{\Phi_F}{\Phi_{F(Q)}} = \frac{k_r + k_i + k_{isc} + k_q[Q]}{k_r + k_i + k_{isc}} = \frac{k_r + k_i + k_{isc}}{k_r + k_i + k_{isc}} + \frac{k_q[Q]}{k_r + k_i + k_{isc}}$$
$$= 1 + k_q[Q]\frac{1}{k_r + k_i + k_{isc}} = 1 + k_q\tau_0[Q] = 1 + K_{SV}[Q] \tag{10.6}$$

where K_{SV} is the Stern–Volmer constant, and τ_0 is the mean fluorescence lifetime of the fluorophore in the absence of quencher. Equation (10.6) is called the Stern–Volmer equation.

Since the fluorescence intensity is proportional to the quantum yield, the Stern–Volmer equation can be written as:

$$\frac{I_0}{I} = 1 + k_q\tau_0[Q] = 1 + K_{SV}[Q] \tag{10.7}$$

where I_0 and I are, respectively, the fluorescence intensities in the absence and presence of quencher (Stern and Volmer 1919). Plotting I_0/I as a function of $[Q]$ yields a linear plot with a slope equal to K_{SV}.

The K_{SV} value shows the importance of fluorophore accessibility to the quencher, while the value of k_q gives an idea of the importance of the diffusion of the quencher within the medium. Figure 10.1 shows a Stern–Volmer plot of fluorescence intensity quenching with iodide of flavin free in solution and of flavin bound to flavocytochrome b_2. The K_{SV} values found are 39 and 14.6 M^{-1} for free and bound flavins, respectively, i.e., values of k_q equal to 8.3×10^9 and 3.33×10^9 M^{-1} s^{-1}, respectively. Accessibility of flavin to KI is more important when it is free in solution, and the presence of protein matrix prevents frequent collisions between iodide and FMN thereby decreasing the fluorophore accessibility to the quencher. Also, as revealed by the k_q values, diffusion of iodide in solution is much more important than in flavocytochrome b_2. The protein matrix inhibits iodide diffusion, thereby decreasing the k_q value.

Figure 10.2 shows a Stern–Volmer plot for the fluorescence intensity quenching by oxygen of zinc protoporphyrin IX embedded in the heme pocket of apomyoglobin ($Mb^{des \rightarrow Fe}$). The slope of the plot yields K_{SV} and k_q values of 15.96 M^{-1} s^{-1} and 7.6×10^9 M^{-1} s^{-1},

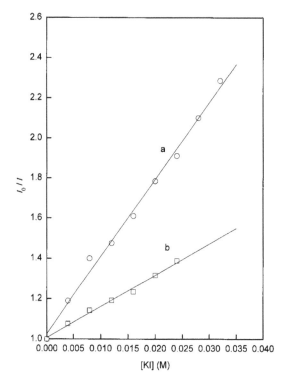

Figure 10.1   Stern–Volmer plots of fluorescence intensity quenching with iodide of FMN free in solution (plot b) and of FMN bound to flavocytochrome b_2 (plot a). Reproduced from Albani, J.R., Sillen, A., Engelborghs, Y. and Gervais, M. (1999). *Photochemistry and Photobiology,* **69**, 22–26, with the permission of the American Society for Photobiology.

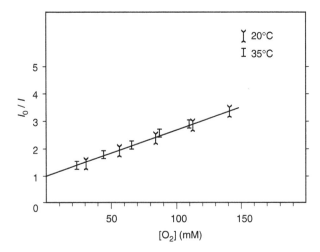

Figure 10.2   Plots of the fluorescence quenching of $Mb^{Fe \rightarrow Zn}$ by oxygen. From the standard deviation of the plots, the same slope was obtained at 20 and 35°C. Source: Albani, J. and Alpert, B. (1987). *European Journal of Biochemistry,* **162**, 175–178 with permission from Blackwell Publishing Ltd.

respectively ($\tau_0 = 2.3$ ns). This k_q value is lower than that (1.2×10^{10} M^{-1} s^{-1}) found for oxygen when fluorescence-quenching of free tryptophan in solution with oxygen is performed. This k_q value is the highest value that can be measured in a homogeneous medium. In the case of Mb$^{Fe \rightarrow Zn}$ (Figure 10.2), the quenching constant, 7.6×10^9 M^{-1} s^{-1}, appears essentially to represent oxygen migration in and near the heme pocket and is temperature-independent. As a result of the short fluorescence lifetime of the zinc porphyrin, $\tau_0 = 2.1$ ns, following dynamic quenching process of zinc-protoporphyrin fluorescence requires oxygen diffusion in spatial proximity to the fluorophore.

A Stern–Volmer plot can be obtained also from fluorescence lifetime quenching. In fact, the fluorescence lifetime in the absence of a quencher is

$$\tau_0 = \frac{1}{k_r + k_i + k_{isc}} \tag{10.8}$$

and in the presence of a quencher, the fluorescence lifetime is

$$\tau_Q = \frac{1}{k_r + k_i + k_{isc} + k_q[Q]} \tag{10.9}$$

$$\frac{\tau_0}{\tau_Q} = 1 + k_q \tau_0 [Q] = 1 + K_{SV}[Q] \tag{10.10}$$

Equations (10.10) and (10.7) are identical. Therefore, in the presence of collisional quenching, we have Equation (10.11) (Figure 10.3):

$$\frac{I_0}{I} = \frac{\tau_0}{\tau_Q} \tag{10.11}$$

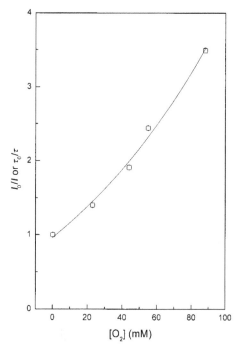

Figure 10.3 Stern–Volmer plot of quenching with oxygen of intensity and lifetime of protoporphyrin IX embedded in the heme pocket of myoglobin. Experiments performed at 20°C.

The most common quenchers are oxygen, acrylamide, iodide, and cesium ions. The k_q value increases with probability of collisions between the fluorophore and quencher. Oxygen is a small and uncharged molecule, so it can diffuse easily. Therefore, the bimolecular diffusion constant k_q observed for oxygen in solution is the most important between all cited quenchers.

Acrylamide is an uncharged polar molecule, so it can diffuse within a protein and quenches fluorescence emission of Trp residues. The quencher should be able to collide with tryptophan whether it is on the surface or in the interior of a protein. Nevertheless, Trp residues, mainly those buried within the core of the protein, are not all reached by acrylamide. For a fully exposed tryptophan residue or for Tryptophan free in solution, the highest value of k_q found with acrylamide is $6.4 \times 10^9 \ M^{-1} \ s^{-1}$.

Cesium and iodide ions quench Trp residues that are present at or near the surface of the protein. The iodide ion is more efficient than the cesium ion, i.e., each collision with the fluorophore induces a decrease in fluorescence intensity and lifetime, which is not the case with cesium. Also, since cesium and iodide ions are charged, their quenching efficiency will depend on the charge of the protein surface. For free tryptophan and tyrosine in solution, the highest values of K_{SV} that we have found with iodide are 17.6 and 19 M^{-1}, respectively, and so the corresponding k_q values are 6.8×10^9 and $5.3 \times 10^9 \ M^{-1} \ s^{-1}$, respectively (Figure 10.4). K_{SV} is calculated from the slope of the plot drawn at low KI concentrations.

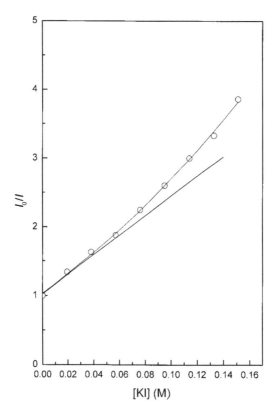

Figure 10.4   Fluorescence intensity quenching of L-Trp by KI. Stock solution $= 4 \ M \ KI + 10^{-3} \ Na_2SO_4$.

10.3 Different Types of Dynamic Quenching

When a protein possesses two or several Trp residues, when quenchers such as iodide, cesium, or acrylamide are used, and if all Trp residues are not accessible to the quencher, the Stern–Volmer equation yields a downward curvature. In this case, we have selective quenching (Figure 10.5b). From the linear part of the plot, we can calculate the value of the Stern–Volmer constant corresponding to the interaction between the quencher and accessible Trp residues. Upon complete denaturation and loss of the tertiary structure of a protein, all Trp residues will be accessible to the quencher. In this case, the Stern–Volmer plot will show an upward curvature. In summary, inhibition of the protein fluorescence with two or several Trp residues can yield three different representations for the Stern–Volmer equations, depending on the accessibility of the fluorophore to the quencher.

1 An exponential plot, i.e., all residues are accessible to the quencher or fluorescence is dominated by one single residue (Figure 10.5a).
2 Downward curvature, i.e., fluorescence is heterogeneous, and residues do not have an identical accessibility to the quencher (Figure 10.5b).
3 Linear plot, i.e., fluorescence is heterogeneous and the accessibility of residues to the inhibitor slightly differs (Eftink and Ghiron 1976).

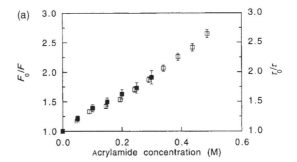

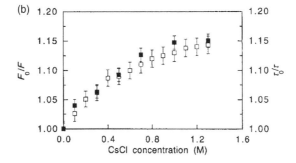

Figure 10.5 Acrylamide and CsCl quenching effects on the ascorbate oxidase emission properties. Steady-state (□) and dynamic (■) quenching of ascorbate oxidase by acrylamide (a) and CsCl (b) upon excitation at 293 nm. Source: Di Venere, A., Mei, G., Gilardi, G. *et al.* (1998). *European Journal of Biochemistry*, **257**, 337–343 with permission from Blackwell Publishing Ltd.

In a selective quencher, the fluorophores fraction accessible to the quencher can be calculated. Let us consider two populations of fluorophore, one accessible to the quencher and the second inaccessible. At high quencher concentrations, fluorescence of the accessible fluorophores is completely quenched. Thus, residual fluorescence originates from inaccessible fluorophores, i.e., fluorophores buried in the hydrophobic core of the protein.

The fluorescence intensity F_0 recorded in the absence of quencher is equal to the sum of the fluorescence intensities of populations accessible (F_a) and inaccessible (F_b) to the quencher:

$$F_0 = F_a + F_b \tag{10.12}$$

In the presence of quencher, only fluorescence of the accessible population decreases according to the Stern–Volmer equation. The fluorescence recorded at a defined quencher concentration $[Q]$ is

$$F = \frac{F_{0(a)}}{1 + K_{SV}[Q]} + F_{0(b)} \tag{10.13}$$

where K_{sv} is the Stern–Volmer constant of the accessible fraction:

$$F_0 - F = \Delta F = \frac{F_{0(a)} \times K_{SV}[Q]}{1 + K_{SV}[Q]} \tag{10.14}$$

$$\frac{F}{\Delta F} = \frac{F_0 + F_0 \times K_{SV}[Q]}{F_{0(a)} \times K_{SV}[Q]} = \frac{F_0}{F_{0(a)} \times K_{SV}[Q]} + \frac{F_0}{F_{0(a)}} \tag{10.15}$$

$$\frac{F_0}{\Delta F} = \frac{1}{f_a \times K_{SV}[Q]} + \frac{1}{f_a} \tag{10.16}$$

where f_a is the fraction of the accessible fluorophore population to the quencher. Therefore, if one plots $F_0/\Delta F$ vs. $1/[Q]$, a linear plot is obtained whose slope is $1/f_a x K_{SV}$, and the intercept is equal to $1/f_a$. Equation (10.16) is known as the Lehrer equation (Lehrer 1971).

Accessibility of a fluorophore to the solvent and thus to the quencher depends on its position within the protein. For example, buried Trp residues should have a lower accessibility to the solvent than those present at the surface.

Figure 10.6 shows fluorescence emission spectra of lens culinaris agglutinin (LCA) (a) ($\lambda_{max} = 330$ nm), of inaccessible Trp residues (b) ($\lambda_{max} = 324$ nm) obtained by extrapolating to $[I^-] = \infty$, and of quenched Trp residues (c) obtained by subtracting spectrum (b) from spectrum (a). The emission maximum of accessible Trp residues is located at 345 nm, a characteristic of emission from Trp residues near the protein surface. Thus, both classes of Trp residues contribute to the fluorescence spectrum of LCA (Albani 1996). The presence of five Trp residues makes the analysis by the modified Stern–Volmer equation very approximate; nevertheless, a selective quenching method allows the percentage of accessible fluorophores to the quencher to be determined.

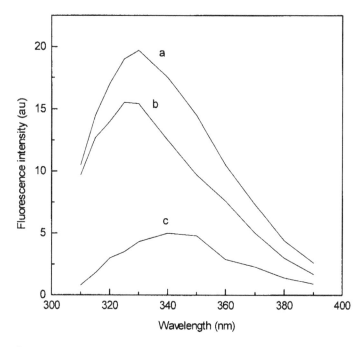

Figure 10.6   Fluorescence emission spectra of LCA (a), of Trp residues exposed to KI (c) and of Trp residues buried in the protein matrix (b). λ_{ex} = 295 nm. Source: Albani, J.R. (1996). *Journal of Fluorescence,* **6**, 199–208, Figure No 5. With kind permission of Springer Science and Business Media (1, 2, and 3).

10.4   Static Quenching

10.4.1   Theory

Fluorescence quenching can also take place by formation in the ground state of a nonfluorescent complex. When this complex absorbs light, it immediately returns to the fundamental state without emitting any photons. This type of complex is called static quenching and can be described by the following equations:

$$F + Q \rightleftarrows FQ \tag{10.17}$$

The association constant K_a is

$$K_a = \frac{[FQ]}{[F]_f[Q]_f} \tag{10.18}$$

where $[FQ]$ is the complex concentration, and $[F]_f$ and $[Q]_f$ are the concentrations of free fluorophore and quencher.

Since the total fluorophore concentration is

$$[F]_0 = [F]_l + [FQ] \tag{10.19}$$

and bound fluorophore does not fluoresce, replacing Equation (10.19) in Equation (10.18) yields

$$K_a = \frac{[F]_0 - [F]_1}{[F]_f[Q]_f}; K_a[Q]_f = \frac{[F]_0 - [F]_f}{[F]_f} = \frac{[F]_0}{[F]_f} - 1$$

$$\frac{[F]_0}{[F]_f} = 1 + K_a[Q]_f \tag{10.20}$$

$$\frac{[I]_0}{[I]} = 1 + K_a[Q]_f \tag{10.21}$$

If we consider the concentration of bound quencher to be very small compared to the added quencher concentration, then the free quencher concentration is almost equal to the total added concentration. Thus, Equation (10.21) can be written as

$$\frac{[I]_0}{[I]} = 1 + K_a[Q] \tag{10.22}$$

Therefore, plotting $[I]_0/[I]$ vs. $[Q]$ yields a linear plot with a slope equal to the association constant of the complex (Figure 10.7).

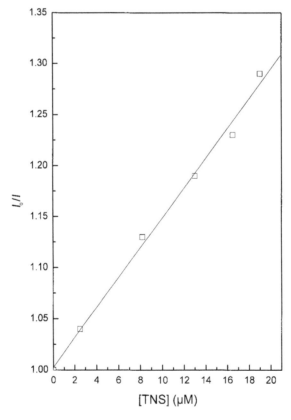

Figure 10.7 Determination of dissociation constant of TNS–LCA complex. $K_a = 0.0148\ \mu M^{-1}$; $K_d = 67\ \mu M$.

Lectins are widely used as tools in the study of plasma membrane modifications in neoplasia. Lectins are able to bind sugars specifically, but the binding constant of the specific free sugar with a lectin may be several orders of magnitude lower than the binding constant of a glycoconjugate containing this sugar. LCA is primarily specific for α-mannopyranosyl residues. The lectin recognizes α-mannopyranosyl end-groups or those substituted at the 0–2 position. Additional requirements for strong binding involve the presence of an L-fucose residue α-1,6-linked to an N-acetylglucosamine (GlcNac) which is linked to the protein via a N-glycosamine bond. This lentil lectin is a mixture of two proteins named LCH-A and LCH-B.

In static quenching, one observes fluorescence of free fluorophore in solution. Complexed fluorophore does not fluoresce and is not observed in the derived equations. The fluorescence lifetime of free fluorophore is the same whether all fluorophore molecules are free or some are complexed. Thus, the fluorescence lifetime of fluorophore does not change upon increasing the quencher concentration. This implies the following equation:

$$\tau_0/\tau_{(Q)} = 1 \tag{10.23}$$

Therefore, in static quenching, one observes an intensity decrease only. Binding of TNS to α_1-acid glycoprotein induces a decrease in the fluorescence intensity of the protein Trp residues. Fluorescence lifetime of the intrinsic fluorophore is not modified. Variations in Trp residue intensities and lifetimes can be analyzed by plotting intensities and lifetimes in the absence and presence of TNS as a function of TNS concentration (Figure 10.8). It is clear from the figure that the TNS–α_1-acid glycoprotein interaction is of a static nature.

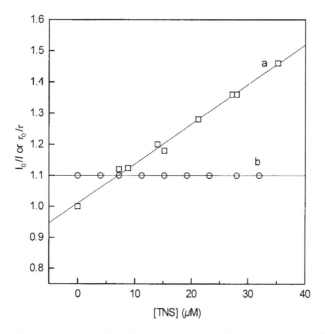

Figure 10.8 Quenching of the tryptophan fluorescence of α_1-acid glycoprotein (4 μM) by TNS at 20°C. Intensity (a) and lifetime (b) variation. Reproduced from Albani, J.R., Sillen, A., Engelborghs, Y. and Gervais, M. (1999). *Photochemistry and Photobiology*, **69**, 22–26, with permission of the American Society for Photobiology.

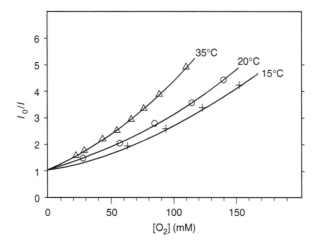

Figure 10.9 Stern–Volmer plots of the fluorescence intensity quenching of Mb^{desFe} by oxygen. The $k\tau_0$ values were obtained from the initial slopes of each plot. Source: Albani, J. and Alpert, B. (1987). *European Journal of Biochemistry*, **162**, 175–178 with permission from Blackwell Publishing Ltd.

In static quenching, an intrinsic or extrinsic fluorophore is used to examine the interaction between two macromolecules.

The fluorophore should be bound to one of the two proteins only (case of extrinsic fluorophores) or should be part of it (case of intrinsic fluorophores). Also, the binding parameters of the fluorophore–macromolecule complex can be determined by following the fluorescence modification of the fluorophore observables.

Finally, one should remember that dynamic quenching is diffusion-dependent, while static quenching is not. Thus, a temperature increase induces an increase in the diffusion coefficient and of k_q (Figure 10.9). However, in the presence of static quenching, increasing the temperature destabilizes the formed complex and thus decreases its association constant (Figure 10.10).

As the slopes of the Stern–Volmer plots are equal to $k\tau_0$, and since the fluorescence lifetimes, τ_0, of both porphyrins do not change with temperature, the slope variation reflects a change that occurs in oxygen diffusion with the temperature. From the slopes at low quencher concentrations, we have found values of k equal to $8.4 \pm 0.4 \times 10^8$ M^{-1} s^{-1} and $13 \pm 0.4 \times 10^8$ M^{-1} s^{-1} at 20° and 35°C, respectively.

Figure 10.10 shows the interaction between TNS and the α_1-acid glycoprotein. We notice that for a constant protein concentration, fluorescence intensity is higher at low temperatures. The K_a obtained is $19.6(\pm4) \times 10^3$ M^{-1}, $16.7(\pm3) \times 10^3$ M^{-1}, and $11(\pm2) \times 10^3$ M^{-1} at 10, 20, and 33°C, respectively. The maximal fluorescence of the complex decreases with increasing temperature ($301 \pm 60, 201 \pm 30$, and 156 ± 30, respectively, in arbitrarily scaled units), as expected from the known temperature dependence of nonradiative decay processes (Albani *et al.* 1995).

Finally, the formation of a complex in the fundamental state can be characterized by disruption of the absorption spectrum of one of the two molecules that forms the complex (Schaberle *et al.* 2003).

Fluorescence quenching is frequently applied to sudy the interaction between humic acids and pollutants. Before presenting one or two applications, we will describe briefly

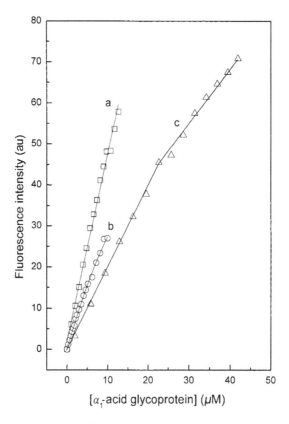

Figure 10.10 Fluorescence intensity of TNS (0.85 μM) in the presence of increased concentration of α_1-acid glycoprotein. (a) = 9°C. (b) = 20°C. (c) = 33.5°C. λ_{ex} = 320 nm and λ_{em} = 430 nm.

some of the terminology used in ecotoxicology. We call bioavailability, fraction of chemical component present in the environmental medium and which is available to be accumulated by the organisms. The environment can include water, sediments, particles in suspension, and food. According to the mode of action and the quantity of pollutant in the medium, two types of toxicity can be observed: acute and chronic toxicities. Acute toxicity describes the adverse effects resulting from a single exposure to a substance (The MSDS HyperGlossary:). Chronic toxicity leads to organisms mortality. Chronic toxicity leads to long-term effects and is evaluated by different types of effects measurements (reproduction, distortions, growth delay...) (Gourlay, 2004).

The hydrophobicity of an organic molecule is quantified by means of its partition coefficient between octanol and water (K_{OW}). K_{OW} is equal to the ratio of compound concentrations at saturation in n-octanol and in water at equilibrium and at a specific temperature. n-Octanol possesses structural properties analogous to lipidic tissues of organisms, and so K_{OW} allows evaluation of the lipophilic character of the contaminant and thus of its capacity to be accumulated in the lipidic tissues of living organisms. The K_{OW} value of hydrophobic organic pollutants (HOP$_s$) is higher than 100. They are nonpolar molecules with a low water solubility. Their weak solubility and their hydrophobicity lead them to bind preferentially on hydrophobic particles of the medium, composed of particulate organic matter (POM) and dissolved organic matter (DOM). POM is organic matter that is retained on a 0.45 μm sieve. It consists of living organisms (phytoplankton, bacteria, and animals) and detritus (e.g., biogenic material in various stages of decomposition).

DOM is composed of small molecules that can be obtained by filtration. Quantification of organic matter in aquatic environment is performed by measuring the concentrations expressed in organic carbon. One measures total organic carbon (TOC) obtained from raw liquid sample, particulate organic carbon (POC) by analyzing the filter, and the dissolved organic carbon (DOC) characterized from the sample after filtration.

Humic substances are the main components of DOM. Thus, the presence of humic substances within the aqueous environment decreases the toxicity level. Organic matter is generally characterized by its absorption located in the range 254–280 nm.

According to their acidity, humic substances are hydrophobic and are split into two groups: humic and fulvic acids. Humic acids are stable molecules originating from the ageing of organic matter. They are responsible for water coloration and represent 40–60% of natural organic matter in rivers and lakes. Fulvic acids are smaller than the humic acids and are generally less aromatic than humic acids extracted from the same pool of DOM.

Soil acidity originates from parental material (quartz, granite, sandstone, etc.), from organic matter (fulvic and humic acids, etc.), and from the pedogenetic evolution. In fact, acidity occurs not only from the presence of H^+ ions but also from Al^{3+} ions. Roots exchange some H^+ ions by Ca^{2+} and Mg^{2+} ions and acidify the medium.

Humic acids are polymerized organic acids of loose structure, which makes them hydrophilic. They are large dark brown or gray-black molecules and are in general rich in oxygen and carbon, and some contain calcium that helps them to bind to clay inducing a humic-clay complex. They are formed in soil by oxidization of the lignin and polyphenols inducing their polymerization. Fulvic acids are poor in carbon and rich in nitrogen, and are found in poorly aerated acidic media. Derived from the Latin *fulvus*, meaning yellow, fulvic acids have the same properties as humic acids but are composed of smaller molecules and are less polymerized.

Humic substance has a complex structure, comprising a hydrophobic core carrying functional radicals, mainly carboxyls and hydroxyls. Thus, humic substances can react with different products of various chemical functions, and so they interact ecologically with all classes of toxic products such as heavy metals, pesticides, etc.

The hydrophobic character of organic pollutants means that they can be adsorbed by sediments or particles in suspension or bind to DOMs such as humic substances. The hydrophobicity of organic pollutants reduces the probability of finding them free in water.

In the presence of very hydrophobic pollutants, three phases can be considered: water, DOM, and POM. Thus, the presence of DOM decreases the bioavailability and thus the harmfulness of the POH_s. Only organic pollutants dissolved and free in water are bioavailable for organisms.

Humic substances that constitute the hydrophobic part of DOM have a high capacity for sorption of hydrophobic organic pollutants. Humic acids have a better sorption capacity than fulvic acids, and thus reduce bioaccumulation much better than fulvic acids.

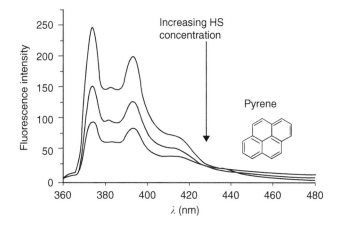

Figure 10.11 Variation of fluorescence emission spectra of pyrene with increasing the humic substance concentration. Source: Frimmel, H.F. (2000). *Agronomie*, **20**, 451–463. Authorization of reprint accorded by INRA, France.

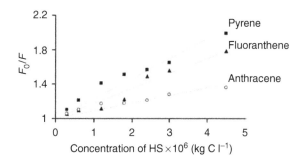

Figure 10.12 Binding constants analysis for quenching of pyrene, fluoranthene, and anthracene fluorescence with Aldrich HS. Reprinted with permission from Perminova, I., Grechishcheva, N. and Petrosyan, V. (1999). *Environmental Science and Technology*, **33**, 3781–3787. Copyright © 1999 American Chemical Society.

Figure 10.11 shows the fluorescence emission spectrum of pyrene in the absence and presence of two humic substance concentrations. The fluorescence intensity of pyrene decreases with increasing humic substance concentration. This decrease results from the binding of pyrene on the humic substance.

Binding of pyrene, fluoranthene, and anthracene to humic substances has been performed by following the decrease in fluorescence intensity of the pollutants in the presence of increased humic substance concentrations. Figure 10.12 shows an analysis of the data based on calculations using Equation (10.22). Since the slopes of the plots are equal to the association constants of the complexes, one can say that pyrene has the highest affinity to humic substances, followed by fluoranthene and anthracene.

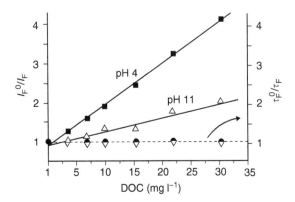

Figure 10.13 Determination of the association constant of a humic substance – pollutants at two pH. Source: Frimmel, H.F. (2000). *Agronomie*, **20**, 451–463. Authorization of reprint accorded by INRA, France.

Binding studies between humic substances and polluants have also been performed by following the variation in fluorescence lifetime for the polluant as a function of increased concentration of humic substances. Binding means static quenching, and so no variation in the fluorescence lifetime should be observed (Figure 10.13).

We can see that while the fluorescence intensity of the polluant is modified in the presence of DOM (DOC), its lifetime does not change. Thus, we have static quenching. Dissolved organic matter–hydrophobic polluant interactions are pH-dependent. In fact, DOC are acids, and so at high pHs, their structure could be modified, and their interaction with hydrophobic polluant would decrease.

10.5 Thermal Intensity Quenching

De-excitation of a fluorophore occurs via different competitive mechanisms described in the Jablonski diagram. The global rate constant, which is the inverse of the fluorescence lifetime, can be considered equal to the sum of the different rates of the competitive mechanisms. Thus, one can write:

$$1/\tau_0 = k_r + k_{isc} + k_i + \sum k_{qi} \tag{10.24}$$

where k_r is the radiative constant, k_{isc} the intersystem crossing constant, k_i the constant corresponding to de-excitation due to the temperature effect, and $\sum k_{qi}$ the different quenching mechanisms such as proton transfer or electron transfer. In the presence of energy transfer at a distance (Förster energy transfer), another rate constant, k_T, should be added.

Working at a constant temperature allows the different parameters of the system to be controlled, and one can quantify the mechanisms intervening in the depopulation of the fluorophore excited state. However, what will happen when we measure the fluorescence lifetime or intensity at different temperatures? Free in solution, a fluorophore can be temperature-dependent or not. In the first case, the temperature variation affects both lifetime and intensity. In the second case, none of these parameters is affected. The temperature increase favors important Brownian motions facilitating energy loss via dynamic quenching by the solvent. This will induce a decrease in the fluorescence quantum yield and in parallel with fluorescence intensity. Fluorescence lifetime is less sensitive to temperature than the quantum yield, and so the observed decrease in fluorescence lifetime with temperature is not proportional to that observed for quantum yield or intensity.

With a temperature increase, k_r decreases. k_{isc} is temperature-independent, and so the temperature will not affect k_r. The constant k_i or thermal constant increases with temperature. k_i is called the solvent constant, since the latter is considered to be responsible for the temperature dependence of the fluorescence lifetime. k_i allows fluorophore activation energy to be determined by the classical Arrhenius theory:

$$k_i = A \exp(-E/RT) \tag{10.25}$$

where A is the temperature-independent factor equal to 10^{15}–10^{17} s^{-1}. E is the Arrhenius activation energy (kcal mol^{-1}), R is the molar gas constant, and T is the temperature in kelvins. Plotting ln k_i as a function of $1/T$ yields the value of E. Taking into consideration that not all constant rates are temperature-dependent, and neglecting the presence of energy transfer, Equation (10.24) can be simplified to

$$1/\tau_0 = k_r + k_i \tag{10.26}$$

Thus, Equation (10.25) can be written as

$$k_i = 1/\tau - k_r = A \exp(-E/RT) \tag{10.27}$$

Since the fluorescence lifetime can be obtained experimentally and k_r calculated, the value of E can be obtained by plotting $\ln(1/\tau - k_r)$ as a function of $1/T$. One should note that since k_i is at least 10 times less than $1/\tau$, it is no longer taken into account in Equation (10.27).

Figure 10.14 displays fluorescence lifetime and quantum yield of FMN with temperature in the absence and presence of AMP. Figure 10.15 shows the Arrhenius plot of free FMN in solution obtained with Equation (10.27).

The slope of fluorescence lifetime variation with temperature of FMN in the absence of AMP differs from that observed in the presence of AMP, thus indicating that an interaction exists between FMN and AMP. The quantum yield of FMN free in solution varies identically to fluorescence lifetime. However, in the presence of AMP, the quantum yield increases, while the lifetime decreases, thus indicating that a static complex (nonfluorescent) is formed between FMN and AMP.

The activation energy of FMN is 0.77 kcal mol^{-1}, a smaller value than that (4.2 kcal mol^{-1}) found for NADH in the same conditions. This clearly means that the difference in the radiotionless transitions is characteristic of the structures of the studied compounds.

The activation energy of the FMN–AMP complex has also been determined and found to be equal to 5.6 kcal mol^{-1}, thus revealing that the structure around the fluorophore and the nature of the interaction between the two molecules will also play an important role in the definition of the activation energy.

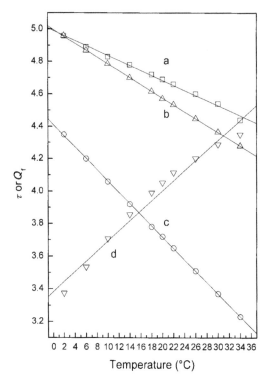

Figure 10.14 Fluorescence lifetime (a and c) and quantum yield (b and d) of free FMN in solution (a and b) and of FMN in the presence of AMP (c and d). The values of the quantum yields are normalized with those of lifetime in order to make a better comparison between the two variations in parameters. Adapted from Spencer, R.D. (1970). Ph.D. thesis, University of Illinois at Urbana Champaign.

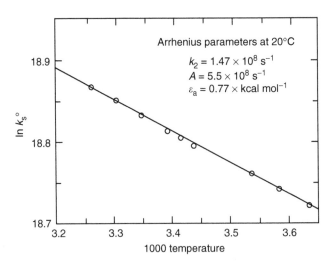

Figure 10.15 Arrhenius plot of the rate of radiationless transitions of FMN in water. Source: Spencer, R.D. (1970). Ph.D. thesis, University of Illinois at Urbana Champaign.

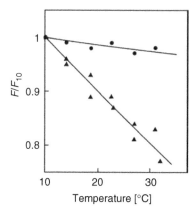

Figure 10.16 Effect of temperature on fluorescence intensity of native (▲) and guanidine unfolded (•) AEDANS-RNase. Intensities are expressed relative to that measured at 10°C for the same sample (native or unfolded). The buffer is 50 mM cacodylate, pH 6.5, in the absence or presence of 6 M guanidine hydrochloride. The protein concentration is 10^{-5} M. Excitation wavelength: 350 nm; emission wavelength: 476 nm. Source: Jullien, M., Garel, J.-R., Merola, F. and Brochon J.-C. (1986). *European Biophysical Journal*, **13**, 131–137, Figure No. 1. With kind permission of Springer Science and Business Media (1, 2, and 3).

The FMN fluorescence is very sensitive to the environment and temperature. In fact, at 34°C, the fluorescence lifetime value of free FMN is 10% lower than that recorded at 2.5°C.

Many other fluorophores are temperature-sensitive only when they are bound to macromolecules. Figure 10.16 shows the effect of temperature on the fluorescence intensity of native and guanidine unfolded AEDANS-RNase. Increasing the temperature from 10 to 30°C induces a decrease in fluorescence intensity for both protein states. The intensity decrease in native protein is more affected by temperature than the guanidine-unfolded protein. This thermal quenching is the consequence of rapid movements of the protein structure around the fluorescent probe. These movements occur during the lifetime of the excited state, and their rate is temperature-dependent.

A fluorophore free in solution shows mainly collisional quenching with the solvent. In proteins, the same fluorophore can be within a ground-state complex or can have important dynamical motions. The second case is observed at temperatures higher than 0°C (Royer *et al.* 1987; Albani 2004).

The temperature variation can affect not only the fluorescence intensity of the spectrum but also its emission bandwidth. However, this is dependent on the fluorophore environment and the fluorophore. Figure 10.17 shows the fluorescence emission spectrum of Trp residues of the protein LCA. In the range of temperatures studied, a shift to the red was not observed, and so we are far from denaturing temperatures. In addition, the emission bandwidth (54 nm) does not change with temperature.

Figure 10.18 shows the fluorescence emission spectrum of TNS bound to LCA recorded at different temperatures. The bandwidth of the spectra increases from 78 to 98 nm with increasing temperature.

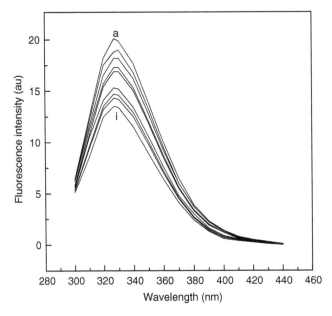

Figure 10.17   Fluorescence spectra of Trp residues of LCA at different temperatures. λ_{ex} = 295 nm. Spectrum a: 6.5°C. Spectrum b: 9.5°C. Spectrum c: 12°C. Spectrum d: 15°C. Spectrum e: 17.8°C. Spectrum f: 20.4°C. Spectrum g: 23°C. Spectrum h: 26°C. Spectrum i: 30°C.

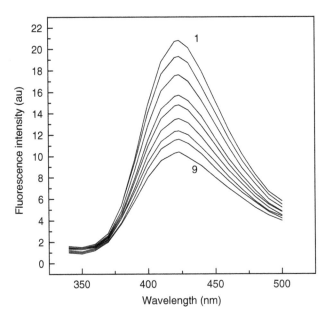

Figure 10.18   Fluorescence spectra of TNS bound to LCA as a function of temperature. λ_{ex} = 320 nm. Spectrum 1: 6.5°C. Spectrum 2: 9.5°C. Spectrum 3: 12°C. Spectrum 4: 15°C. Spectrum 5: 17.8°C. Spectrum 6: 20.4°C. Spectrum 7: 23°C. Spectrum 8: 26°C. Spectrum 9: 30°C.

References

Albani, J. and Alpert, B. (1987). Fluctuation domains in myoglobin. Fluorescence quenching studies. *European Journal of Biochemistry*, **162**, 175–178.

Albani, J., Vos, R., Willaert, K. and Engelborghs, Y. (1995). Interaction between human α_1-acid glycoprotein (orosomucoid) and 2-*p*-toluidinylnaphthalene-6-sulfonate. *Photochemistry and Photobiology*, **62**, 30–34.

Albani, J.R. (1996). Dynamics of *Lens culinaris* agglutinin studied by red-edge excitation spectra and anisotropy measurements of 2-*p*-toluidinylnaphthalene-6-sulfonate (TNS) and of tryptophan residues. *Journal of Fluorescence*, **6**, 199–208.

Albani, J.R. (2004). Effect of the secondary structure of carbohydrate residues of α_1-acid glycoprotein (orosomucoid) on the local dynamics of Trp residues. *Chemistry and Biodiversity*, **1**, 152–160.

Albani, J.R., Sillen, A., Engelborghs, Y. and Gervais, M. (1999). Dynamics of flavin in flavocytochrome b_2: a fluorescence study. *Photochemistry and Photobiology*, **69**, 22–26.

Di Venere, A., Mei, G., Gilardi, G. *et al.* (1998). Resolution of the heterogeneous fluorescence in multi-tryptophan proteins: ascorbate oxidase. *European Journal of Biochemistry*, **257**, 337–343.

Eftink, M.R. and Ghiron, C.A. (1976). Exposure of tryptophanyl residues in proteins. Quantitative determination by fluorescence quenching studies. *Biochemistry*, **15**, 672–680

Frimmel, H.F. (2000). Development in aquatic humic chemistry. *Agronomie*, **20**, 451–463.

Gourlay, C. (2004). Biodisponibilité des hydrocarbures polycycliques dans les écosystèmes aquatiques: influence de la matière organique naturelle et anthropique. Thèse de doctorat à l'Ecole Nationale du Génie Rural, des Eaux et Forêts Centre de Paris.

Jullien, M., Garel, J-R., Merola, F. and Brochon J.-C. (1986). Quenching by acrylamide and temperature of a fluorescent probe attached to the active site of ribonuclease. *European Biophysical Journal*, **13**, 131–137.

Lehrer, S.S. (1971). Solute perturbation of protein fluorescence. The quenching of the tryptophyl fluorescence of model compounds and of lysozyme by iodide ion. *Biochemistry*, **10**, 3254–3263

Perminova, I., Grechishcheva, N. and Petrosyan, V. (1999). Relationships between structure and binding affinity of humic substances for polycyclic aromatic hydrocarbons: relevance of molecular descriptors. *Environmental Science and Technology*, **33**, 3781–3787.

Royer, C.A., Tauc, P., Hervé, G. and Brochon, J.-C. (1987). Ligand binding and protein dynamics: a fluorescence depolarization study of aspartate transcarbamylase from *Escherichia coli*. *Biochemistry*, **26**, 6472–6478.

Schaberle, F.A., Kuz'min, V.A. and Borissevitch, I.E. (2003). Spectroscopic studies of the interaction of bichromophoric cyanine dyes with DNA. Effect of ionic strength. *Biochimica Biophysica Acta*, **1621**, 183–191.

Spencer, R.D. (1970). Fluorescence lifetimes: theory, instrumentation and application of nanosecond fluorometry. Ph.D. thesis, University of Illinois at Urbana Champaign. Published by University Microfilms International, Ann Arbor, MI.

Stern, O. and Volmer, M. (1919). Uber die Abklingungszeit der Fluoreszenz. *Physikalische Zeitschrift*, **20**, 183–188.

Chapter 11
Fluorescence Polarization

11.1 Definition

Biological systems show global rotation and local dynamics that are dependent on the structure, environment, and function of the system. These motions differ from one system to another, and within one system, local motions are not the same. The best-known example is that of membrane phospholipids where the hydrophilic phosphates are rigid, and the hydrophobic lipid is highly mobile. Polarized light is a good tool for studying different types of rotations a molecule can undergo.

Natural light is unpolarized; it has no preferential direction. Thus, in order to study molecules dynamics during the excited-state lifetime, molecular photoselection should be performed. This can be reached by exciting the fluorophores with a polarized light and by recording the emitted light in a polarized system. When excitation is performed with polarized light (definite orientation), absorption of the fluorophore will depend on the orientation of its dipole in the ground state compared to the polarized excitation light (Figure 11.1).

Fluorophores with dipoles perpendicular to excitation light will not absorb. Fluorophores with dipoles parallel to excitation light will absorb the most. Thus, polarized excitation will induce photoselection in the fluorophore absorption. The electric vector of excitation light is oriented parallel to or in the same direction as the z-axis. Emitted light will be measured with a polarizer. When emission is parallel to the excitation, the measured intensity is called $I_\parallel$. When the emission is perpendicular to the excitation light, the measured intensity is called $I_\perp$. Light polarization and anisotropy are obtained according to Equations (11.1) and (11.2):

$$P = \frac{I_\parallel - I_\perp}{I_\parallel + GI_\perp} \tag{11.1}$$

and

$$A = \frac{I_\parallel - I_\perp}{I_\parallel + 2GI_\perp} \tag{11.2}$$

G is the correction factor and allows one to take into account the differences in sensitivity of the detection system in the two polarizing directions I_{vv} and I_{vh}. The G factor can be

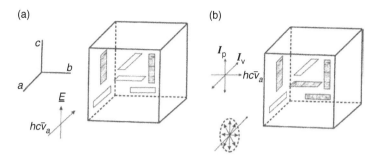

Figure 11.1 Photoselection of differently oriented molecules with plane-polarized (a) and unpolarized (b) light. Excited molecules are shaded. After Albrecht, A.C. (1961). *Journal of Molecular Spectroscopy*, **6**, 84. Figure published by Dörr, F. (1971), *Creation and Detection of the Excited States*, pp. 53–122, Dekker, New York.

evaluated by measuring the fluorescence intensities parallel and perpendicular to the z-axis with the excitation polarizer perpendicular to the z-axis. G will be equal to:

$$G = \frac{I_{vh}}{I_{hh}} \tag{11.3}$$

P and A are inter-related with the following two equations:

$$P = \frac{3A}{2 + A} \tag{11.4}$$

and

$$A = \frac{2P}{3 - P} \tag{11.5}$$

Since the presence of polarizers induces photoselection, the global fluorescence intensity recorded in the presence of polarizers is lower than that obtained in their absence. This is a consequence of the decrease in the number of absorbing and emitting fluorophores.

A fluorophore free in solution can have a low polarization value, whereas when it is bound to a macromolecule, its polarization increases. The polarization unit is a dimensionless entity, i.e., the value of P does not depend on the intensity of emitted light and the fluorophore concentration. However, this is the theory; the reality is quite different. In fact, measuring polarization at high fluorophore concentrations yields erroneous values, and in many cases, instead of reading the correct values of P and A, values that neighbor the limiting values are recorded.

Equation (11.1) indicates that values of P occur between 1 ($I_\perp = 0$) and -1 ($I_\parallel = 0$). Natural or unpolarized light, where $I_\perp = I_\parallel$, yields a P value of 0. These two extreme values of P are observed when the polarized absorption transition moment and that of the emission are colinear ($I_\perp = 0$) or perpendicular ($I_\parallel = 0$). However, in a rigid medium where motions are absent, the absorption and emission transition dipoles can be oriented by an angle θ, one relative to the other.

Therefore, the probability of light absorption by the molecules will be a function of $\cos^2 \theta$, and the polarization will be equal to

$$P_o = \frac{3 \cos^2 \theta - 1}{\cos^2 \theta + 3} \tag{11.6}$$

and

$$A_o = \frac{3 \cos^2 \theta - 1}{5} \tag{11.7}$$

θ	P_o	A_o
0	0.5	0.4
45	0.14	0.1
54.7	0.00	0.00
90	−0.33	−0.20

A value of $\theta = 0$ indicates that the excitation and emission dipoles are co-planar. This means that excitation induces an $S_0 \rightarrow S_1$ transition. For an angle of 54.7°, P and A are 0. This angle is called the "magic angle." By measuring fluorescence emission with polarizers set up with an orientation equal to the magic angle, unpolarized light can be detected. P_o and A_o are called, respectively, intrinsic polarization and anisotropy. P_o and A_o are measured at temperatures equal to $-45°C$ or less. At this temperature, fluorophores do not show any movement. Colinear excitation and emission vectors yield $S_0 \rightarrow S_1$ and $S_1 \rightarrow S_0$ transitions. This means that high positive values of P would correspond to an $S_0 \rightarrow S$, transition, while negative values of P would correspond to the $S_0 \rightarrow S_2$ transition.

Since electronic transitions differ from one excitation wavelength to another, the value of P would change with excitation wavelength. Emission generally occurs from the lowest excited state $S_1 \nu_0$, and so one can measure anisotropy or polarization along the absorption spectrum at a fixed emission wavelength. We obtain a spectrum called the excitation polarization spectrum or simply the polarization spectrum (Figure 11.2).

Figure 11.2 shows the excitation polarization spectrum of protoporphyrin IX in propylene glycol at $-55°C$ (full line) and bound to the heme pocket of apohemoglobin recorded at 20°C (dotted line). One can see that polarization at a low temperature is higher than that observed when porphyrin is embedded in the heme pocket of apohemoglobin. This is the result of fluorophore local motions within the pocket, independently of the global rotation of the protein.

Finally, it is important to mention that the polarization spectrum shape is characteristic of the chemical structure and nature of the fluorophore.

11.2 Fluorescence Depolarization

11.2.1 *Principles and applications*

In vitrified solution and/or when fluorophore molecules do not show any residual motions, the measured polarization is equal to the intrinsic one, P_o. This value is obtained at low

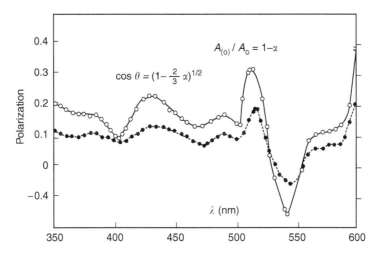

Figure 11.2 Excitation polarization spectrum of protoporphyrin IX in propylene glycol at −55°C (full line) and bound to the heme pocket of apohemoglobin recorded at 20°C (dotted line). λ_{em} = 634 nm. Source: Sebban P. (1979). Etude par fluorescence de molécule Hb^{desFe} thesis, University of Paris 7.

and ambient temperatures, too, if, during the excited-state lifetime of the fluorophore, i.e., during the time period between light absorption and emission, the orientation of excitation and emission dipoles does not change.

However, during the excited-state lifetime, energy transfer to neighboring molecules and/or local and global motions of the fluorophores can be observed. These two phenomena induce reorientation of the emission dipole, thereby depolarizing the system. Therefore, the value of the measured polarization P will be lower than intrinsic polarization value P_0 (Weber, 1952).

The relationship between P and P_0 is given by the Perrin equation:

$$\frac{1}{P} - \frac{1}{3} = \left(\frac{1}{P_0} - \frac{1}{3}\right)\left(1 + \frac{RT\tau_0}{\eta V}\right) = \left(\frac{1}{P_0} - \frac{1}{3}\right)\left(1 + \frac{\tau_0}{\phi_R}\right) \tag{11.8}$$

where R is the gas constant = 2.10 kcal mol^{-1} deg^{-1} or 0.8×10^3 g cm^2 s^{-2} M^{-1} deg^{-1}; T, the temperature in kelvins; η the medium viscosity in centipoise (cp) or g cm^{-1} s^{-1} (1 cp = 0.01 g cm^{-1} s^{-1}); and V and ϕ_R the rotational volume and rotational correlation time of the fluorophore.

Plotting $1/P$ as a function of T/η yields in principle a linear graph with a y-intercept equal to $1/P_0$ and a slope equal to

$$\frac{R\tau_0}{V}\left(\frac{1}{P_0} - \frac{1}{3}\right)$$

The Perrin plot can also be written as

$$\frac{1}{A} = \frac{1}{A_0}\left(1 + \frac{\tau_0}{\phi_R}\right) \tag{11.9}$$

Decreasing the temperature or increasing the concentration of sucrose and/or glycerol in the medium increases the medium viscosity.

Whether fluorophores are intrinsic or extrinsic to the macromolecule (protein, peptide, or DNA), depolarization is the result of two motions, fluorophore local motions and macromolecule global rotation.

Rotational correlation time of a spherical macromolecule can be determined using Equation (11.10) or Equation (11.11):

$$\phi_P = M(v + h)\eta/kTN \tag{11.10}$$

where M is the molecular mass, $v = 0.73$ cm^3 g^{-1} characterizes the specific volume, $h = 0.3$ cm^3 g^{-1} is the hydration degree, η is the medium viscosity, and N is the Avogadro number.

$$\phi_P(T) = 3.8\eta(T) \times 10^{-4} M \tag{11.11}$$

Equations (11.10) and (11.11) do not yield exactly the same value for ϕ_P. For example, for a protein with a molecular mass of 235 kDa, such as flavocytochrome b$_2$ extracted from the yeast *Hansenula anomala*, the rotational correlation times calculated from Equations (11.10) and (11.11) are 97 and 90 ns, respectively.

The Perrin plot enables us to obtain information concerning fluorophore motion. When the fluorophore is tightly bound to the protein, its motion will correspond to that of the protein. In this case, ϕ_R will be equal to the protein rotational correlation time ϕ_P, and A_o obtained experimentally from the Perrin plot will be equal to that measured at $-45°C$ (Figure 11.3).

The rotational correlation times measured at 15°C and 35°C are equal to 5.5 and 3.6 ns. These values, within the range of theoretical values of myoglobin, indicate that zinc porphyrin is tightly linked to the myoglobin heme pocket. The extrapolated value of P is 0.176. This value, identical to that (0.1744) measured for the polarization P_o at $-20°C$, does not reveal the presence of motions. Thus, Perrin plot experiments show that zinc-protoporphyrin IX is rigid within the heme pocket. This result is in good agreement with that found by Raman spectroscopy. In fact, the authors found that the zinc–imidazole bond is anomalously weak, orientation of the proximal histidine is suboptimal for binding to the zinc, and it is held relatively rigid in the myoglobin heme pocket (Andres and Atassi 1970).

When fluorophore exhibits local motions, extrapolated value of A, $A(o)$, is lower than the A_o value obtained at $-45°C$. Information obtained from the Perrin plot slope depends on the fluorophore fluorescence lifetime and on the relative amplitudes of the fluorophore and protein motions.

Equation (11.9) indicates the possibility of calculating the rotational correlation time of the fluorophore not only by varying the T/η ratio but also by adding a collisional quencher. Interaction between the quencher and fluorophore decreases the fluorescence lifetime and intensity of the fluorophore, and increases its fluorescence anisotropy. Plotting $1/A$ as a function of τ_o yields a straight line with a slope equal to ϕ_R. If the fluorophore is tightly bound to the macromolecule and does not exhibit any residual motions, the measured ϕ_R is equal to ϕ_P, and the extrapolated anisotropy is equal to that measured at a low temperature.

If, however, the fluorophore exhibits free motion, the measured ϕ_R is lower than that of the protein, and extrapolated anisotropy is lower than the limiting anisotropy.

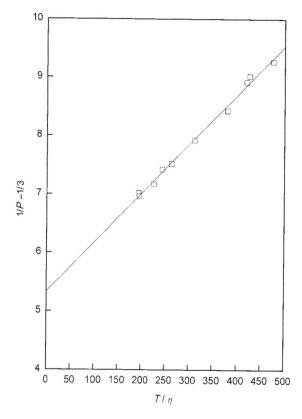

Figure 11.3 Steady-state fluorescence polarization vs. temperature/viscosity ratio for $Mb^{Fe \rightarrow Zn}$. Data were obtained by thermal variation of temperature. $\lambda_{ex} = 517$ nm and $\lambda_{em} = 600$ nm.

11.3   Fluorescence Anisotropy Decay Time

Anisotropy measurements yield information on molecular motions taking place during the fluorescence lifetime. Thus, measuring the time-dependent decay of fluorescence anisotropy provides information regarding rotational and diffusive motions of macromolecules (Wahl and Weber, 1967). Time-resolved anisotropy is determined by placing polarizers in the excitation and emission channels, and measuring the fluorescence decay of the parallel and perpendicular components of the emission.

In the time-decay method, fluorescence anisotropy can be calculated as follows:

$$A(t) = D(t)/S(t) \tag{11.12}$$

where

$$D(t) = V_v = gV_H \tag{11.13}$$

$$S(t) = V_v + 2gV_H \tag{11.14}$$

V_v and V_H are fluorescence intensities of vertically and horizontally polarized components, respectively, obtained with vertically polarized light. g is the correction factor and is equal to H_V/H_H, where H_V and H_H are the vertically and horizontally polarized components, respectively, of the fluorescence elicited by horizontally polarized light.

For a fluorophore bound tightly to a protein, anisotropy decays as a single exponential,

$$A(t) = A_o e^{-t}\Phi_P \tag{11.15}$$

where Φ_P is the rotational correlation time of the protein, and A_o is the fluorophore intrinsic anisotropy at the excitation wavelength.

When a fluorophore exhibits segmental motions, time-resolved anisotropy decay must be analyzed as the sum of exponential decays:

$$A(t) = A_o[\alpha e^{-t/\phi_S} + (1-\alpha)e^{-t/\phi_L}] \tag{11.16}$$

where ϕ_S and ϕ_L are the short and long rotational correlation times, α and $1 - \alpha$ are the weighting factors for the respective depolarizing processes.

$$\frac{1}{\phi_S} = \frac{1}{\phi_P} + \frac{1}{6\phi_F} \tag{11.17}$$

and

$$\frac{1}{\phi_L} = \frac{1}{\phi_P} \tag{11.18}$$

Φ_F is the rotational correlation time of the fluorophore.

11.4   Depolarization and Energy Transfer

Energy transfer is a source of depolarization. For example, a high energy-transfer efficiency in hemoproteins can be evidenced by plotting Perrin plot at different temperatures.

The cytochrome b_2 core from the yeast *Hansenula anomala* has a molecular mass of 14 kDa, and its sequence shows the presence of two tryptophan residues. Their fluorescence intensity decay can be adequately described by a sum of three exponentials. Lifetimes obtained from the fitting are equal to 0.054, 0.529, and 2.042 ns, with fractional intensities equal to 0.922, 0.068, and 0.010. The mean fluorescence lifetime, τ_o, is 0.0473 ns.

The main fluorescence lifetime ($\tau = 54$ ps) and its important fractional intensity ($f_i = 92\%$) indicate that an important energy transfer occurs between Trp residues and heme. In an attempt to measure rotational correlation time of the protein, we have measured anisotropy of cytochrome b_2 core Trp residues at different temperatures. Results are described with the classical Perrin plot ($1/A$ as a function of T/η) (Figure 11.4).

The data yield a rotational correlation time equal to 38 ps instead of 5.9 ns calculated theoretically for the cytochrome b_2 core, with an extrapolated value $A(o)$ of 0.208, lower than that (0.265) usually found for Trp residues at $\lambda_{ex} = 300$ nm at $-45°C$. The fact that the extrapolated anisotropy is lower than the limiting anisotropy means that the system is depolarized as a result of global and local motions within the protein. In this case, the value of the apparent rotational correlation time (Φ_A) calculated from the Perrin plot is lower than the global rotational time of the protein (Φ_P). However, the fact that Φ_A is 1000 times lower than Φ_P indicates that a third process different than the global and local rotations is

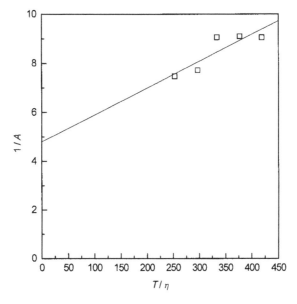

Figure 11.4 Steady-state fluorescence anisotropy vs. temperature/viscosity ratio for tryptophan residues of cytochrome b_2 core. Data are obtained by thermal variations in the range 10–36°C.

participating in the depolarization of the system. This process is the high energy transfer that occurs from the tryptophans to the heme.

The extrapolated value $A(o)$ is

$$A(o) = A_o \times d_P \times d_F \times d_T \tag{11.19}$$

where A_o, d_P, d_F, and d_T are the intrinsic anisotropy, depolarization factor due to global protein rotation, depolarization factor due to the local fluorophore motions, and depolarization factor due to energy transfer and Brownian motion, respectively.

In the absence of energy transfer, Equation (11.19) becomes

$$A(o) = A_o \times d_P \times d_F \tag{11.20}$$

References

Albrecht, A.C. (1961). Polarizations and assignments of transitions: The method of photoselection. *Journal of Molecular Spectroscopy*, **6**, 84.

Andres, S.F. and Atassi, M.Z. (1970). Conformational studies on modified proteins and peptides. Artificial myoglobins prepared with modified and metalloporphyrins. *Biochemistry*, **9**, 2268–2275.

Dörr, F. (1971). Polarized light in spectroscopy and photochemistry, in: A.A. Lamola (ed.), *Creation and Detection of the Excited States*, pp. 53–122, Dekker, New York.

Sebban P. (1979). Etude par fluorescence de molécule Hb[desFe] thesis, University of Paris 7.

Wahl, P. and Weber, G. (1967). Fluorescence depolarization of rabbit gamma globulin conjugates. *Journal of Molecular Biology*, **30**, 371–382.

Weber, G. (1952). Polarization of the fluorescence of macromolecules. I. Theory and experimental method. *Biochemical Journal*, **51**, 145–155.

Chapter 12
Interaction Between Ethidium Bromide and DNA

12.1 Objective of the Experiment

This experiment allows students to learn how to extract and purify DNA and then to study the interaction between purified DNA and a fluorescent probe such as ethidium bromide (3,8-diamino-5-ethyl-6-phenylphenanthridinium bromide).

Ethidium bromide is a common fluorescent stain for nucleic acids. It is reported to have significant anti-tumor and anti-viral properties. However, its potential applications in human therapy are prevented, due to its mutagenic and carcinogenic activities in model organisms.

Students will follow the absorption and fluorescence spectral modifications of ethidium bromide in the presence of different DNA concentrations. Then, they will calculate the number of binding sites and the mean association constant. Before coming to the lab, students should determine the absorption and emission spectra of ethidium bromide bound to DNA.

12.2 DNA Extraction from Calf Thymus or Herring Sperm

12.2.1 Destruction of cellular structure

About 1 g of tissue is placed onto a glass strip over crushed ice and carefully cut in all directions with a razor blade. The tissue is then transferred into a 50 ml polypropylene tube Falcon sunk in crushed ice. Ten milliliters of TE buffer (i.e., TRIS buffer 10 mM pH 7.6; EDTA 1 mM) is added, and the cells are carefully homogenized with a glass rod for 5 min. SDS solution [50 p.100 in Ethanol 50 p.100 in water (v/v)] is added to yield 1% in SDS as a final concentration, and the mixture is gently stirred for 5 min at room temperature with the glass rod in order to avoid frothing.

12.2.2 DNA extraction

One volume of chloroform/2-methyl-1-butanol (3/1 ; v/v) is added and the tube is carefully blocked and violently shaken for 10 min at room temperature. Then, the volume in the

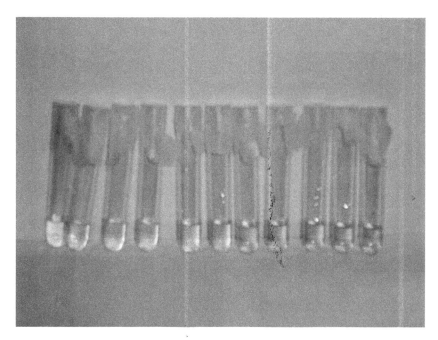

Color plate 6.2 Methylnitrophenol dissolved in 10 mM Tris buffer at pHs ranging from 2 (far left) to 12 (far right). The colorless protonated methylnitrophenol turns yellow when the pH increases.

Color plate 12.4 Fluorescence cuvettes containing ethidium bromide (left) and ethidium bromide–DNA complex (right) solutions.

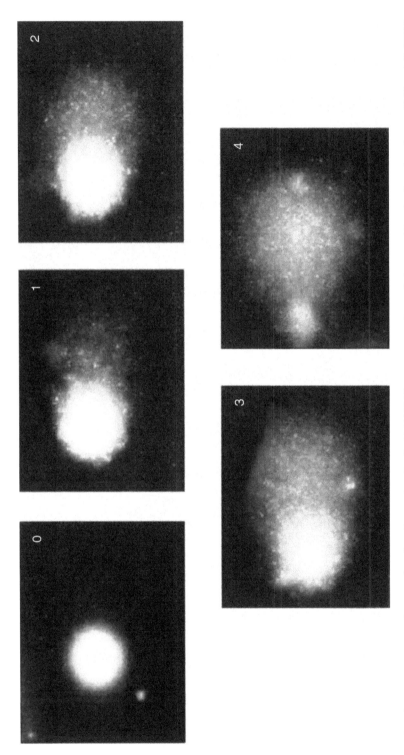

Color plate 16.7 Images of comets (from lymphocytes), stained with DAPI. These represent classes 0–4, as used for visual scoring. Source: Collins, A.R. (2004). *Molecular Biotechnology*, **26**, 249–261. Reprinted with permission from Humana Press Inc.

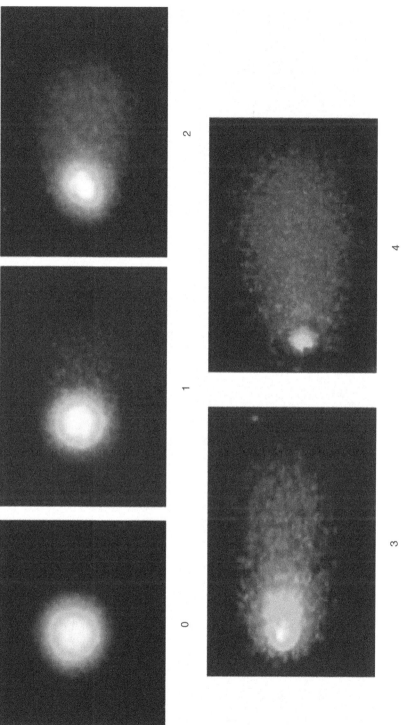

Color plate 16.8 CHO cells treated with ethyl methanesulfonate; Magnification 350×. Courtesy of Dr. Alok Dhawan, Industrial Toxicology Research Centre, Mahatma Gandhi Marg, Lucknow, UP. www.cometassayindia.org and www.cometassayindia.org/protocols.htm.

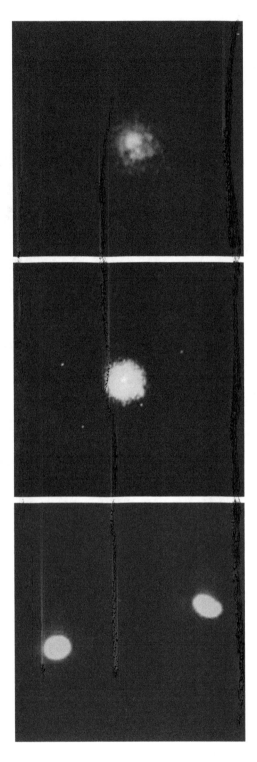

Color plate 16.9 Comet test for lymphocyte cells in three states: normal, apoptotic, and necrotic. The reader can refer to the following site: www.cometassayindia.org and www.cometassayindia.org/protocols.htm.

tube is made up to 40 ml with 2 M NaCl (this improves the solubilization of DNA in the hydrophilic phase) before being regularly shaken for a further 5 min.

Centrifugation is performed for 10 min at 3500 rpm. The upper phase is then removed and precipitated by 1 volume of ethanol at –20°C. DNA fibers are simply coiled on a glass rod and dissolved immediately in the TE buffer. DNA fibers cannot support the dehydrating effect of ethanol for more than 40 s.

12.2.3 DNA purification

Proteases [DNase free] are added to the DNA solution to decrease the amount of protein that can remain associated. After the protease action, the entire procedure of DNA extraction is repeated, and the DNA dissolved again in TE buffer and dialyzed against TE buffer to eliminate any proteases and peptides. Professors and students can also refer to the following website for DNA extraction and purification: http://gmotraining.jrc.it/docs/Session04.pdf

12.2.4 Absorption spectrum of DNA

Plot an absorption spectrum of DNA solution from 220 to 300 nm. Add 5 μl of the DNA stock to 1 ml buffer, then plot the spectrum and calculate the concentration from the optical density (OD) at 260 nm. The ε value at 260 nm for herring DNA is 12 858 $M^{-1}(bp)\ cm^{-1}$ and for calf thymus DNA 6900 $M^{-1}(bp)\ cm^{-1}$.

To check the DNA purity, a comparison of the optical densities at 260 and 280 nm should be performed. The OD_{260}/OD_{280} ratio should give a value of around 1.8 for pure DNA.

12.3 Ethidium Bromide Titration with Herring DNA

Prepare an ethidium bromide solution (50 μM) in a fluorescence cuvette and then plot both absorption and emission spectra ($\lambda_{ex} = 470$ nm). The molar extinction coefficient of ethidium bromide at 480 nm is 5680 $M^{-1}\ cm^{-1}$.

To the fluorophore solution, add 8–9 μM DNA aliquots, and after each addition plot both absorption and emission spectra. The experiment will end after 18 DNA additions.

Take two photos, one of ethidium bromide before adding DNA solution, the second at the end of the experiment.

1 What are the indications showing that ethidium bromide is binding to DNA?
2 Compare the spectra obtained with those you find from the literature.
3 Measure the fluorescence intensity at the peak, correct it for dilution, then for the inner filter effect, and plot the corrected fluorescence intensity as a function of added DNA concentration. Use the data to determine the number of binding sites and the association constant of the complex.

4   From the absorption spectrum of bound DNA you find in the literature, can you tell if there are any wavelengths where bound ethidium bromide does not absorb? Also, from the absorption spectra you recorded, are there any wavelengths where bound ethidium bromide does not absorb or its absorption is weak? Plot at this wavelength the OD of ethidium bromide as a function of added DNA concentration. Use the data to determine the number of binding sites and the association constant of the complex. Compare the results obtained in 3 and 4.

12.4   Results Obtained with Herring DNA

12.4.1   *Absorption and emission spectra*

The OD at 260 nm of herring DNA is found to be equal to 0.115 (see absorption spectrum, Figure 12.1). This yields a concentration equal to 9.1 μM in the cuvette (for each 5 μl taken from the stock and added to 1 ml of Tris buffer). The stock concentration is

$$9.1\ \mu M \times 1005\ /\ 5 = 1829\ \mu M \tag{12.1}$$

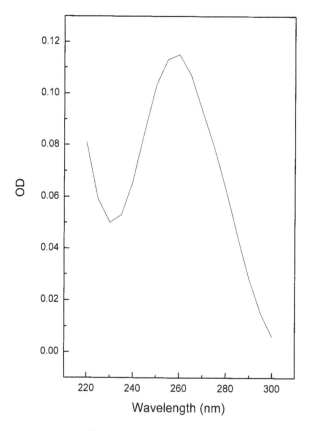

Figure 12.1   Absorption spectrum of herring DNA.

The OD at 280 nm is 0.063. Thus, the ratio of the ODs, OD_{260}/OD_{280}, is 1.825, a value very close to 1.8, indicating that DNA is pure.

In general, DNA contamination with protein can be calculated using the following equation:

$$[Protein] = (1.55 \times OD_{280}) - (0.76 \times OD_{260})$$ (12.2)

Equation (12.2) is to be applied only when the ratio of the optical densities is lower than 1.8.

The fluorescence intensity of ethidium bromide increases in the presence of DNA (Figure 12.2). Once saturation is reached, the intensity increase stops. This intensity increase means that complex formation is occurring. The intensity increase is proportional to the number of binding sites and to the affinity constant.

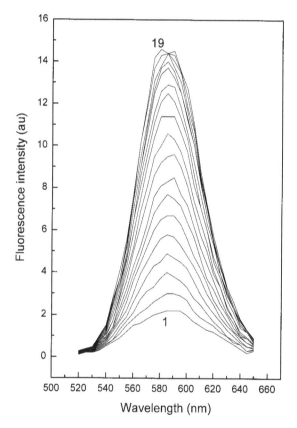

Figure 12.2 Fluorescence emission spectra of 50 μM ethidium bromide in the absence (1) and presence (2–19) of herring DNA (9.1–164 μM). λ_{ex} = 470 nm.

Upon binding of ethidium bromide to DNA, the maximum OD from its absorption spectrum decreases, together with a shift to higher wavelengths (Figure 12.3). An isobestic point is observed at 510–511 nm, indicating that the fluorophore molecules are in two spectrophotometrically different conditions, bound and free. This feature characterizes the interaction of ethidium bromide to DNA. The isobestic point at 512 nm is generally observed when an interaction occurs between ethidium bromide and double-stranded DNA. In the presence of a single-stranded DNA, the same type of shift is observed with an isobestic point located at 507 nm. The absorption decrease at 480 nm indicates that a number of ethidium bromide molecules free in the solution decrease in the presence of DNA. The absorption of ethidium bromide bound to DNA is located at 520 nm. Thus, the absorption spectra show clearly that ethidium bromide free in solution shows an absorption spectrum different from that of bound fluorophore.

Finally, one can see that upon addition of DNA to the ethidium bromide solution, the orange color of the fluorophore turns to pink (Figure 12.4). This color modification is the result of the interaction between the two molecules and mainly binding of ethidium bromide to DNA.

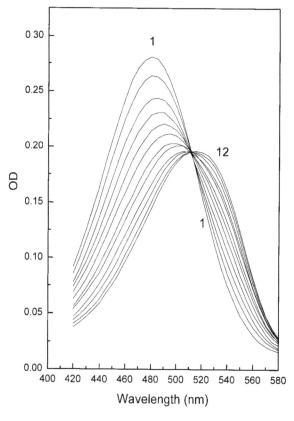

Figure 12.3 Absorption spectra of 50 μM ethidium bromide in the absence (1) and presence (2–12) of DNA. Spectrum 12 corresponds to 100 μM of DNA. An isobestic point is observed at 511 nm, indicating the presence of two species for ethidium bromide free and bound to DNA.

Figure 12.4 Fluorescence cuvettes containing ethidium bromide (left) and ethidium bromide–DNA complex (right) solutions. Reproduced in Color plate 12.4.

12.4.2 Analysis and interpretation of the results

The results obtained are identical to many data obtained by different authors. However, one should remember that the literature also shows a fluorescence spectrum of bound ethidium bromide located at 605 nm to the difference of the peak observed for free ethidium bromide (see, e.g. the fluorescence excitation and emission spectra published by Molecular Probe) (Figure 12.5). However, this is not always observed, as is the case here.

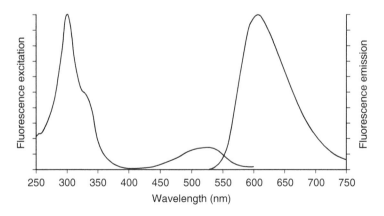

Figure 12.5 Fluorescence excitation and emission spectra of ethidium bromide–DNA complex published by Molecular Probe.

In general, when titration experiments are performed, it is preferable to keep the fluorophore concentration constant and to increase the macromolecule concentration (protein, DNA, etc.). Titration of the fluorophore will induce an increase or decrease in fluorescence intensity. The shape of the titration curve allows the binding parameters of the complex to be calculated. In our case, we observe an increase in fluorescence intensity (Figures 12.2 and 12.6) and a decrease in the OD (Figures 12.3 and 12.7). The fluorescence titration curve (Figure 12.6) reaches a plateau, which allows determining number of binding sites. The number of binding sites is equal to $120/50 = 2.5$ sites. The dissociation constant is equal to the concentration corresponding to a 50% increase in fluorescence intensity. This corresponds to a concentration of 45.5 μM or an association constant of 0.22×10^5 M^{-1}. Since we have more than one binding site, the calculated binding constant is a mean one.

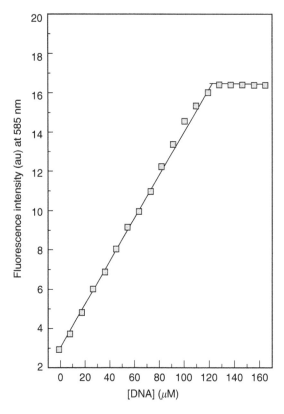

Figure 12.6 Titration of 50 μM ethidium bromide with herring DNA. $\lambda_{ex} = 470$ nm. The number of binding sites is equal to $120/50 = 2.5$ sites and the mean dissociation constant is equal to 45.5 μM, i.e., an association constant equal to $K_a = 0.22 \times 10^5$ M^{-1}.

The excitation spectrum of bound ethidium bromide published by Molecular Probes (Figure 12.5) indicates that, at saturation, fluorophore in the presence of DNA does not absorb in the region between 400 and 420 nm. Also, we can see from the absorption spectra of ethidium bromide recorded at different concentrations of DNA (Figure 12.3) that the OD at 420 nm of the free probe decreases in the presence of increasing concentrations of DNA. Therefore, by plotting OD at 420 nm as a function of added DNA, we can obtain the number of binding sites of ethidium bromide on DNA and the binding constant of the complex. We find 2.3 sites with a mean dissociation constant equal to 60 μM, i.e., an association constant equal to $K_a = 0.167 \times 10^5$ M^{-1} (Figure 12.7).

The results obtained from absorption and fluorescence data are identical, showing that both methods can be used to characterize DNA and ethidium bromide interaction.

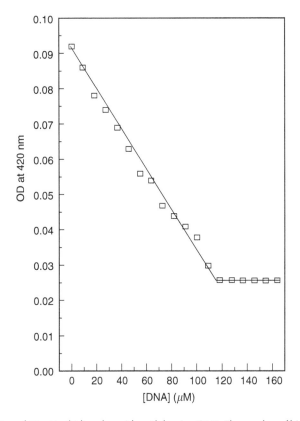

Figure 12.7   Titration of 50 μM ethidium bromide with herring DNA. The number of binding sites is equal to $115/50 = 2.3$ sites, and the mean dissociation constant is equal to 60 μM, i.e., an association constant equal to $K_a = 0.167 \times 10^5$ M^{-1}.

Data for the absorption spectra can also be analyzed by the following method.

Let us consider a solution of a product A with concentration equal to $[A_o]$. The OD of A measured at a specific wavelength λ is OD$_{(o)} = \varepsilon$cl. Let us add to the A solution increasing concentrations of a B solution. After each addition, the OD at λ is measured.

If the product B and the AB complex do not absorb at λ, the decrease in OD observed at λ will characterize the decrease in concentration of free A.

Thus, one can describe the interaction between A and B with the following equation:

$$A + B \Leftrightarrow AB \tag{12.3}$$

The association constant of the complex is

$$K_a = [AB]/[A_f][B_f] \tag{12.4}$$

where $[A_f]$ and $[B_f]$ are the concentrations of free A and B compounds:

$$K_a[B_f] = \frac{[AB]}{[A_f]} = \frac{[A_o] - [A_f]}{[A_f]} = \frac{[A_o]}{[A_f]} - 1 \tag{12.5}$$

Equation (12.5) can be written as:

$$\frac{[A_o]}{[A_f]} = 1 + K_a[B_f] \tag{12.6}$$

Since the OD measured at λ is proportional to $[A_f]$, Equation (11.6) can be written as

$$\frac{OD_o}{OD} = 1 + K_a[B_f] \tag{12.7}$$

If we consider the concentration of bound B to be very small compared to that of added B, then the free B concentration is almost equal to the total added concentration. Thus, Equation (12.7) can be written as

$$\frac{OD_o}{OD} = 1 + K_a[B] \tag{12.8}$$

Therefore, plotting $OD_o/OD_{obs} = f([B])$ yields a linear plot whose slope is equal to the association constant of the complex. This method of analysis can be applied to the interaction between DNA and ethidium bromide. In fact, by plotting OD ratios at 420 nm with Equation (13.8), one can obtain the graph shown in Figure 12.8. The association constant K_a calculated from the slope of the plot is 0.1×10^5 M^{-1}. This value is in the same range of those obtained in Figures 12.6 and 12.7.

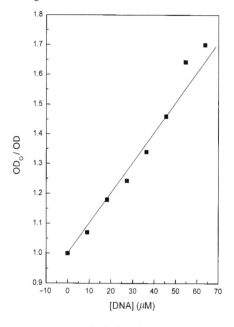

Figure 12.8 Plot of the OD ratio at 420 nm of ethidium bromide as a function of DNA concentration.

Application of Equation (13.8) is correct when the ligand (here ethidium bromide) is very small compared to the macromolecule (DNA) and when the ligand binds almost completely when it is added to the macromolecule. This was observed also for the interaction between hemin and apomyoglobin or apocytochrome, and between TNS and almost all proteins.

12.5   Polarization Measurements

Titration of a fixed amount of ethidium bromide (40 μM) with DNA is performed independently from the other experiments. The polarization of ethidium bromide increases upon DNA addition. Once the stoichiometry of the complex is reached, a plateau is observed (Figure 12.9).

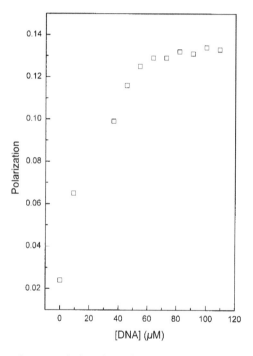

Figure 12.9   Polarization of 40 μM ethidium bromide in the presence of herring DNA. $\lambda_{ex} = 520$ nm and $\lambda_{em} = 580$ nm.

Polarization values are calculated manually by measuring the emission intensities in the presence of the polarizers at the excitation and the emission wavelengths. Thus, it is possible to calculate the total emission intensity at each DNA concentration. The plot of I_F vs. [DNA] concentration (Figure 12.10) yields a result identical to that observed with the polarization or when intensities were measured without polarizations (Figure 12.6).

In parallel to the polarization experiment, we measure ODs at 420 nm after each DNA addition. Also, the results (Figure 12.11) are identical to those already described in Figure 12.7. In the three plots, we have an ethidium bromide:DNA stoichiometry of 2:1.

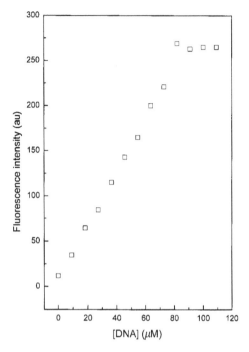

Figure 12.10   Fluorescence emission intensity of 40 μM ethidium bromide with DNA concentration. Data are obtained from the polarization experiment.

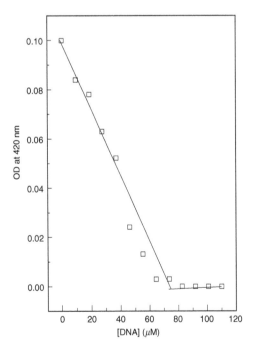

Figure 12.11   OD of 40 μM ethidium bromide with DNA.

One should be careful in measuring polarization, since if the sample is not fluorescing too much, errors in reading the intensities could be frequent. Also, depending on the instrument the student is using to perform the measurements, experiments could be easy or very hard to conduct. Therefore, it is important that all students perform each experiment described here so that they can learn how to conduct all experiments, difficult and easy, and also compare all the data obtained together.

12.6 Results Obtained with Calf Thymus DNA

The same studies were performed with calf thymus DNA. In addition to the absorption and emission spectra, we have recorded the fluorescence excitation spectra of ethidium bromide at different DNA concentrations (Figure 12.12). One can see that the increase in fluorescence intensity stops when the stoichiometry of the complex is reached. Binding parameters ($n = 2.6$ and $K_a = 3 \times 10^5$ M^{-1}) determined from the fluorescence excitation intensities are found to be equal to those calculated with the fluorescence emission and the OD variations. Figure 12.13 shows the normalized intensity increase in both fluorescence emission and excitation intensity peaks of ethidium bromide with the function of DNA. The intensity increase in the excitation spectrum of ethidium bromide in the presence of DNA is the result of binding of the fluorophore to DNA and thus of the increase in the number of excited bound fluorophores.

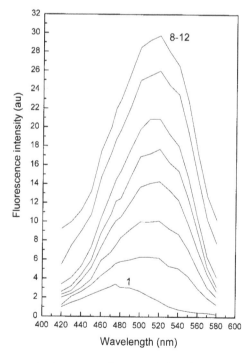

Figure 12.12 Fluorescence excitation spectra of 23 μM ethidium bromide (spectrum 1) with increasing concentrations of calf thymus DNA (spectra 2–12) ($\lambda_{em} = 600$ nm).

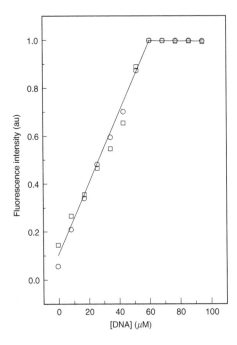

Figure 12.13   Intensity variation, excitation (circles) (λ_{em} = 600 nm) and emission (squares) (λ_{ex} = 470 nm) of 23 μM of ethidium bromide with DNA concentration.

12.7   Temperature Effect on Fluorescence of the Ethidium Bromide–DNA Complex

DNA is stable at high temperatures up to 60–65°C. Raising the temperature induces DNA denaturation, and the two-single strands split up. This can be evidenced by following the OD variation of DNA at 260 nm at increasing temperatures from 20 to 82°C (Marmur and Doty 1959) (see also Figure 12.14).

We suggest here following DNA denaturation by following the fluorescence emission and variation of excitation spectra of the ethidium bromide–DNA complex at increasing temperatures from 50 to 85°C. This experiment should be done at the end of a titration experiment when you have complete binding of ethidium bromide within DNA strands. Figure 12.15 shows clearly that the fluorescence intensities of both excitation and emission spectra decrease when the temperature increases. It is important to indicate that free ethidium bromide in solution does not show any significant fluorescence.

Plotting the normalized fluorescence intensities of excitation and emission spectra vs temperature yields a sigmoidal plot (Figure 12.16) with an inflection point equal to that observed when thermal DNA denaturation is observed by following the ODs at 260 nm (Figure 12.14).

The melting temperature, T_m, of DNA is equal to that when the two macromolecule strands split. If we consider that upon complete denaturation, the fluorescence intensity is 0.6, T_m can be calculated at half denaturation at 0.8 and so is 70.5 $\pm$ 0.5°C. This value is equal to that found when OD is measured as a function of temperature. One

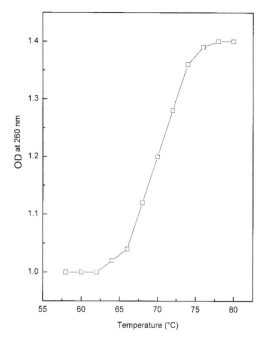

Figure 12.14 Thermal denaturation of DNA followed at an OD of 260 nm.

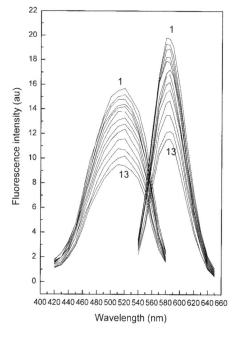

Figure 12.15 Fluorescence excitation (left) (λ_{em} = 600 nm) and emission (right) (λ_{ex} = 520 nm) spectra of ethidium bromide–DNA complex with temperature. Temperatures from spectra 1 to 13 are: 51, 55, 59.5, 62.1, 64.5, 67.1, 68.5, 70.9, 72.8, 74, 76.6, 80.8, and 84°C.

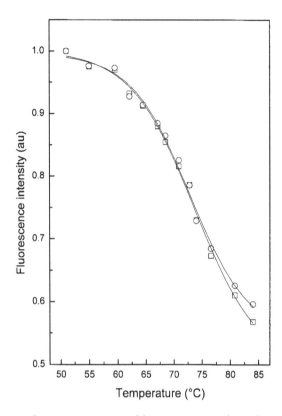

Figure 12.16   Variation in fluorescence intensity of the BET–DNA complex with temperature. The data are from three experiments. □: emission intensity at 585 nm $\lambda_{ex} = 520$ nm. ○: excitation intensity at 520 nm ($\lambda_{em} = 600$ nm).

should expect to see a release of ethidium bromide from the denatured DNA, but this is not the case since denatured DNA shows a statistical coil structure where ethidium bromide can still bind. However, one should not exclude the possibility of some of the ethidium bromide being free in solution with DNA denaturation.

References

Duhamel, J., Kanyo, J., Dinter-Gottlieb, G. and Lu, P. (1996). Fluorescence emission of ethidium bromide intercalated in defined DNA duplexes: Evaluation of hydrodynamics components. *Biochemistry*, **35**, 16687–16697.

Leng, F., Chaires, J.B. and Waring, M.J. (2003). Energetics of echinomycin binding to DNA. *Nucleic Acids Research*, **31**, 6191–6197.

Luedtke, N.W., Liu, Q. and Tor, Y. (2005). On the electronic structure of ethidium. *Chemistry – A European Journal*, **11**, 495–508.

Marmur, J. and Doty, P. (1959). Heterogeneity in deoxyribonucleic acids. I. Dependence on composition of the configurational stability of deoxyribonucleic acids. *Nature*, **183**, 1427–1429.

Shafer, R.H., Brown, S.C., Delbarre, A. and Wade, D. (1984). Binding of ethidium and bis(methidium)spermine to Z DNA by intercalation. *Nucleic Acids Research*, **12**, 4679–4690.

Somma M. (2004). Extraction and purification of DNA. In: Querci M, Jermini M, and Van den Eede, G. (eds), *The Analysis of Food Samples for the Presence of Genetically, Modified Organisms*. European commission, Joint research centre, special publication 1.03.114, edition.

Song, G., Li, L., Liu, L., Fang, G., Lu, S., He, Z. and Zeng, Y. (2002). Fluorometric determination of DNA using a new ruthenium complex $Ru(bpy)_2PIP(V)$ as a nucleic acid probe. *Analytical Sciences*, **18**, 757–759.

Vardevanyan, P.O., Antonyan, A.P., Parsadanyan, M.A., Davtyan, H.G. and Karapetyan, K.T. (2003). The binding of ethidium bromide with DNA: interaction with single- and double-stranded structures. *Experimental and Molecular Medicine*, **35**, 527–533.

Wahl, Ph., Paoletti, J. and Le Pecq, J.-B. (1970). Decay of fluorescence emission anisotropy of the ethidium bromide–DNA complex evidence for an internal motion in DNA. *Proceedings of the National Academy of Sciences*, USA, **65**, 417–421.

Chapter 13
Lens culinaris Agglutinin: Dynamics and Binding Studies

Students should perform a set of three experiments. The best way to perform these experiments is to have different groups working separately on the three types of experiments. This needs at least three persons present to supervise the experiments and to help the students when necessary. Each experiment can be conducted during half a day or one day. The last day should be used to regroup the results and to discuss the data.

13.1 Experiment 1. Studies on the Accessibility of I^- to a Fluorophore: Quenching of Fluorescein Fluorescence with KI

13.1.1 Objective of the experiment

The following experiment aims to help students perform a dynamic quenching experiment and to find out the role of the fluorophore micro-environment in the modification of the dynamic constants.

Before entering the laboratory, students should be able to differentiate between dynamic and static quenching.

Students will perform two different quenching experiments using KI as quencher, one with fluorescein free in solution and the second with fluorescein bound covalently to the protein *Lens culinaris* agglutinin (LCA). The protein is primarily specific for α-mannopyranosyl residues. The lectin recognizes α-mannopyranosyl end-groups or those substituted at the 0–2 position. Additional requirements for strong binding involve the presence of an L-fucose residue α-1,6-linked to an N-acetylglucosamine (GlcNac-l) which is linked to the protein via a N-glycosamine bond.

13.1.2 Experiment

Students should receive a stock solution of 4 M KI in the presence of 0.02 M of $Na_2S_2O_3$ to avoid the formation of I_3^{-1}.

At 495 nm, the extinction molar coefficient (ε) of free fluorescein in water and of fluorescein covalently bound to LCA is 42 and 132.996 mM^{-1}cm^{-1}, respectively. Students should not forget to correct the observed fluorescence intensities from fluorescein for the dilution and, if necessary, for the inner filter effect (λ_{ex} = 480 nm and λ_{em} = 515 nm).

a   In the fluorescence cuvette containing 1 ml of PBS buffer, add with a Pasteur pipette 5 grains of fluorescein from Sigma. Measure the optical density (OD) at 495 nm (to obtain the concentration of the fluorescein in the cuvette) and at 480 and 515 nm (to be sure that you are working in good conditions). The ODs at the excitation and emission wavelengths should not exceed 0.01. Once your fluorescein solution is ready, plot the fluorescence emission spectrum from 500 to 560 nm.

Then, add 6 aliquots of 10 μl of KI each, mix slowly, and plot the fluorescence spectrum after each addition. Repeat the experiments two to three times.

How does the fluorescence emission spectrum of fluorescein vary with the addition of KI? Explain.

b   Measure the intensity of each spectrum at the peak and correct it for the dilution. Then, draw the Stern–Volmer plot for each experiment using the corrected intensities. Measure the Stern–Volmer constant, K_{SV}, and the bimolecular diffusion constant, k_q, from each experiment. Finally, calculate the mean values of the three experiments. The value of the fluorescence lifetime of free fluorescein at the excitation wavelength is 4 ns.

c   Repeat the preceding experiment with the LCA–fluorescein complex instead of free fluorescein. Determine the values of K_{SV} and k_q, and compare the results with those obtained with free fluorescein in buffer. The value of the fluorescence lifetime of fluorescein–LCA complex at the excitation wavelength is 3.16 ns. What do you conclude? Explain.

If students have access to a fluorescence lifetime instrument, it would be useful to see how one can measure fluorescence lifetime. In this case, it will be useful if students can perform the experiments described by following fluorescence lifetime quenching with KI and compare their results with intensity quenching experiments.

13.1.3   Results

a   Figure 13.1 shows the fluorescence emission spectra of fluorescein free in buffer in the presence of increasing concentrations of KI. We observe a decrease in the intensity of the fluorescence emission spectrum of fluorescein in the presence of KI (Figure 13.1). The fluorophore and quencher diffuse in solution, and after each collision a decrease in the fluorescence intensity of the fluorophore occurs.

b and c   Dynamic fluorescence quenching of FITC with iodide analyzed by the Stern–Volmer equation yields a K_{SV} equal to 9.608 ± 0.273 and 3.795 ± 0.295 M^{-1}, for fluorescein free in buffer (Figure 13.2a) and for that bound to LCA (Figure 13.2b), respectively. Thus, iodide is accessible to bound FITC on LCA. However, this accessibility is lower than that observed for free fluorescein in solution, because amino acid residues surrounding the probe decrease the frequency of collisions with the quencher. The bimolecular quenching constant of iodide is 2.402 ± 0.068 and $1.160 \pm 0.090 \times 10^9$ M^{-1} s^{-1} when the interaction occurs with free

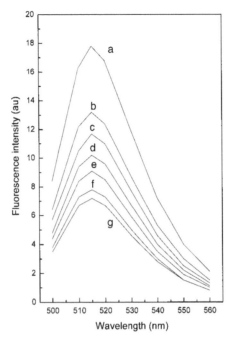

Figure 13.1   Fluorescence emission spectra of fluorescein free in buffer (a) in presence of increasing concentrations of KI (spectra b to g). [KI] at g is equal to 0.135 M.

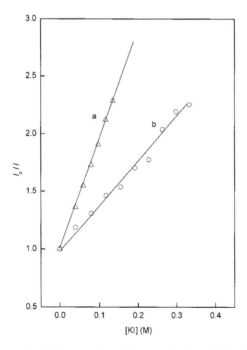

Figure 13.2   Stern–Volmer plot for the quenching of fluorescein (a) and of the fluorescein–LCA complex (b) with KI. $\lambda_{ex} = 480$ nm and $\lambda_{em} = 515$ nm.

fluorescein and fluorescein bound to LCA, respectively. Thus, diffusion of iodide in the vicinity of bound fluorescein is hindered by the surrounding amino acid residues.

13.2 Experiment 2. Measurement of Rotational Correlation Time of Fluorescein Bound to LCA with Polarization Studies

13.2.1 Objective of the work

The purpose of these experiments is first to learn how to measure polarization value of a fluorophore and then to see the difference in the information when we perform the experiments at different temperatures and when we work at a fixed temperature but increasing the concentration of sucrose in the medium.

A good model for these studies is a fluorophore bound covalently to a protein; this is why we choose the fluorescein–LCA complex.

13.2.2 Polarization studies as a function of temperature

a   Calculate at 20°C the rotational correlation time of the lectin with one of these two following formulae:

$$\phi_P = M(v + h)\eta/kTN \tag{11.10}$$

where M is the molecular mass $= 49\,000$, $v = 0.73 \text{ cm}^3 \text{ g}^{-1}$ characterizes the specific volume, $h = 0.3 \text{ cm}^3 \text{ g}^{-1}$ is the hydration degree, η is the medium viscosity, and N is the Avogadro number.

$$\phi_P(T) = 3.8\eta(T) \times 10^{-4} \text{ M} \tag{11.11}$$

b   Measure the polarization of the fluorescein–LCA complex at different temperatures from 5 to 35°C, every 5° ($\lambda_{ex} = 495$ nm and $\lambda_{em} = 515$ nm). Plot the Perrin plot using Table 13.1.

What are the values of the rotational correlation time at 20°C and of the extrapolated polarization? Compare the value of ϕ_P with that calculated theoretically in (a) and the extrapolated polarization with that measured at −65°C and which is 0.452 ($\lambda_{ex} = 495$ nm).

Do you obtain the same values? What is the significance of your data? Explain your results, and what information you can obtain from them?

13.2.3 Polarization studies as a function of sucrose at 20° C

c   Observe the changes in polarization of the fluorescein–LCA solution at a constant temperature as successive additions of sucrose alter the viscosity. A table of sucrose aliquots and the corresponding temperature over viscosity terms (T/η) is provided below.

Table 13.1 Values of water viscosities at different temperatures.

$t°C$	η	$t°C$	η	$t°C$	η	$t°C$	η
0	1.794	11	1.274	21	0.986	31	0.787
1	1.732	12	1.239	22	0.963	32	0.771
2	1.674	13	1.206	23	0.941	33	0.755
3	1.619	14	1.175	24	0.918	34	0.740
4	1.568	15	1.145	25	0.898	35	0.735
5	1.519	16	1.116	26	0.878	36	0.710
6	1.473	17	1.088	27	0.858	37	0.697
7	1.429	18	1.062	28	0.839	38	0.684
8	1.387	19	1.036	29	0.821	39	0.671
9	1.348	20	1.011	30	0.803	40	0.659
10	1.310						

Source: Tuma, J. J. (1983). *Handbook of Physical Calculations*, McGraw-Hill, New York.

Prepare a sucrose solution at 60%. Dissolve 12 g of sucrose in 10 ml of buffer by heating and mixing. Once the sucrose is completely dissolved, complete with buffer up to 20 ml.

Add 165 μl of the sucrose solution to 835 μl of the protein solution; the result is a solution with 10% sucrose. The volume of the solution is 1 ml. Measure the polarization of the solution.

Add 200 μl of the sucrose solution to the 1 ml of the protein solution. The new solution is now at 20% sucrose. New volume: 1200 μl. Measure the polarization of the solution.

Add 110 μl of the sucrose solution to the 1200 μl of the protein solution. The new solution is now at 25% sucrose. New volume: 1310 μl. Measure the polarization of the solution.

Add 120 μl. The new solution is now at 30% sucrose. New volume: 1430 μl. Measure the polarization of the solution.

Add 130 μl. The new solution is now at 35% sucrose. New volume: 1560 μl. Measure the polarization of the solution.

Add 145 μl. The new solution is now at 40% sucrose. New volume: 1705 μl. Measure the polarization of the solution.

Add 345 μl. The new solution is now at 50% sucrose. New volume: 2050 μl. Measure the polarization of the solution.

Add 410 μl. The new solution is now at 60% sucrose. New volume: 2460 μl. Measure the polarization of the solution.

The values of the viscosity at 20°C for the different concentrations of sucrose are:

%	0	10	20	25	30	35	40	50	60
η	1.002	1.337	1.945	2.447	3.107	4.023	0.107	15.43	50.49
T/η	292	219	150	120	92	68	47.5	19	5

Plot the Perrin plot. What type of plot do you obtain and what information do you obtain?

13.2.4   *Results*

a   Theoretically rotational correlation time is 20 ns.

b   The Perrin plot obtained by varying the temperature yields a rotational correlation time equal to 2.9 ns at 20°C and an extrapolated value $P(0)$ equal to 0.390. The extrapolated value is much lower than the limiting polarization measured at −65°C (0.452). This means clearly that fluorescein displays free local motions independent of that of the lectin. This is confirmed by the fact that the rotational correlation time calculated from the slope of Perrin plot is much lower than the rotational correlation time of the lectin. Thus, it is an apparent correlation time due to the global rotation of the lectin and the local motion of the fluorescein.

The Perrin plot as a function of temperature (Figure 13.3) reveals the presence or absence of local motions of the fluorophore on the macromolecule. The absence of any residual motions yields a Perrin plot with a slope equal to that of the protein and an extrapolated polarization equal to the limiting one.

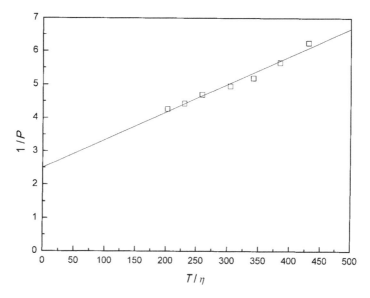

Figure 13.3   Perrin plot of the fluorescein–LCA complex obtained at different temperatures. $\lambda_{ex} = 495$ nm and $\lambda_{em} = 515$ nm. $\Phi_A = 2.9$ ns. $1/P(0) = 2.56$; $P(0) = 0.390$.

c   It is possible to obtain experimentally the rotational correlation time of the fluorophore by increasing the medium viscosity in the presence of glycerol and/or sucrose. In this case, the global rotation of the protein is completely hindered, and only local motions of the fluorophore are observed. $1/P$ as a function of T/η yields a plot with two slopes. Rotational correlation times of the protein (ϕ_P) and of the fluorophore (ϕ_R) are calculated at a high and low T/η, respectively (Figure 13.4).

The slope at high T/η values gives a ϕ_P of 18 ns, a value in the same range (20 ns) of that calculated theoretically. The other slope yields a Φ_R of 72 ps; this value characterizes the free rotation of a fluorophore and which is generally found to be equal to 100 ps.

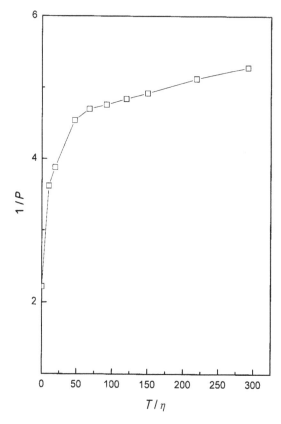

Figure 13.4 Perrin plot of the LCA–fluorescein complex at 20°C and at different sucrose concentrations. The slope of the high values gives a Φ_P of 18 ns and the other slope yields a Φ_R of 72 ps.

13.3 Experiment 3. Role of α-L-fucose in the Stability of Lectin–Glycoproteins Complexes

13.3.1 Introduction

Serotransferrins (STF) from blood plasma and lactotransferrins (LTF) from mammalian milk are glycoproteins that bind to LCA. Besides their role in iron transport and the inhibition of the growth of micro-organisms, the human sero- and lactotransferrin share the following common properties: (1) their molecular mass is around 76 kDa; (2) they are composed of a single polypeptide chain of 679 and 691 amino acid residues for STF and LTF, respectively, organized in two lobes originating from a gene duplication; (3) each lobe binds reversibly one Fe^{3+} ion; (4) the protein moiety presents a high degree of homology (about 62%); (5) they are glycosylated (6.4% by weight).

The two lobes correspond to the N-terminal and C-terminal halves of the molecules and are tightly associated by non-covalent interactions. Also, both are joined by a connecting short peptide of 12 and 11 amino acids in sero- and lactotransferrin, respectively. The three-dimensional pictures of the two proteins are perfectly superimposable, with very few differences.

Human STF contains two carbohydrates of the *N*-acetyllactosaminic type, located in the C-terminal lobe of the polypeptide chain. The two glycosylation sites (Asn-413 and 611) may be occupied by bi-, tri-, and tetra-antennary carbohydrates.

Carbohydrates of human LTF are located in both N- and C-domains, at three glycosylation sites (Asn-137, 478 and 624).

Carbohydrates of human STF are not fucosylated, while those of human LTF have an α-1,6-fucose bound to the *N*-acetyl glucosamine residue linked to the peptide chain, and an α-1,3-fucose bound to the *N*-acetyllactosamine residues.

Studies have shown that the α-1,3-fucose residue does not play any role in the interaction between LCA and glycopeptides isolated from human LTF or chemically synthesized (Kornfeld *et al.* 1981). However, the same work showed that α-1,6-fucose is important for a tight binding. Also, X-ray diffraction studies have proved that the α-1,6 fucose is essential for attaining the proper binding conformation of the carbohydrates (Bourne *et al.* 1993).

13.3.2   *Binding studies*

The goal of these studies is to measure the association constants of the LTF–LCA and STF–LCA complexes. LCA is given as a fluorescein–LCA complex (from Sigma). The extinction coefficient of the complex is 132.996 mM^{-1} cm^{-1} at 495 nm. Two molecules of fluorescein are bound to one molecule of LCA. Binding of the STF or LTF to LCA–fluorescein complex induces a variation in the fluorescence intensity of the extrinsic probe. LTF and STF concentrations can be determined spectrophotometrically at 280 nm with an absorption $E^{1\%} = 14.3$ and 14.0, respectively.

Prepare stock solutions of LTF and STF equal to 28 μM.

a   Prepare in a fluorescence cuvette around 0.8 μM of LCA–fluorescein solution in PBS buffer (final volume equal to 1 ml). Then, plot the fluorescence emission spectrum from 500 to 560 nm, $\lambda_{ex} = 495$.

Add 5 μl aliquots of LTF or STF solution to the fluorescence cuvette and mix slowly. Then, plot the fluorescence emission spectrum from 500 to 560 nm.

Continue the experiments by adding 10 aliquots of 5 μl each of the LTF or STF solution.

Repeat each experiment twice.

How does the emission intensity of the LCA–fluorescein complex vary with the addition of LTF or STF? Explain.

b   Why do we need to use an extrinsic probe to perform our experiments instead of adding LTF or STF to LCA and following the fluorescence intensity variation of Trp residues of LCA?

c   Measure the intensity of each spectrum at the peak (515 nm) and correct it for the dilution. Then, plot the corrected volume as a function of STF or LTF concentration.

From the shape of the plots drawn, what can you say about the affinities of the transferrins to LCA?

d The dissociation constants can be calculated using the following equation:

$$\Delta F/F = (\Delta F_{max}/F)(L_b/P) \tag{13.1}$$

where ΔF, ΔF_{max}, F, L_b, and P are the fluorescence change for a concentration L of the glycoprotein, the maximum fluorescence change at saturation of the protein with the glycoprotein, the fluorescence intensity of LCA–FITC in the absence of glycoprotein, the concentration of bound glycoprotein, and the total concentration of LCA–FITC, respectively. The concentration of bound glycoprotein can be calculated from the root of the quadratic equation [Equation (13.2)] arising from the definition of the binding constant:

$$L_b = 0.5\{(P_0 + L_0 + K_d) - [(P_0 + L_0 + K_d)^2 - 4P_0L_0]^{1/2}\} \tag{13.2}$$

where P_0 is the protein concentration.

Replacing Equation (13.2) in Equation (13.1) gives Equation (3):

$$\Delta F/F = (\Delta F_{max}/F) \times \{(P_0 + L_0 + K_d + [(P_0 + L_0 + K_d)^2 - 4PL]^{1/2}\}/2P_0 \tag{13.3}$$

ΔF_{max} can be obtained by plotting $1/\Delta F$ as a function of $1/L_0$.

From the two plots you obtain, is it possible to determine ΔF_{max} for both LCA–LTF and LCA–STF interactions? Explain.

Note: the whole demonstration of Equation (13.3) is given in Chapter 15.

e After you have determined ΔF_{max}, find the value of the dissociation constants by applying Equation (13.3). You can do this manually by choosing two to three values of ΔF and L_0. For each set of ΔF and L_0, you will have a value for K_d; then you calculate the mean value of K_d. What is your conclusion concerning the affinities?

f Calculate the values of the affinity constants with Equation (10.22). Do you find the same values as those obtained with Equation (13.3). Can you explain?

13.3.3 Results

a How does the emission intensity of the LCA–fluorescein complex vary with the addition of LTF or STF? Explain.

Here, we are studying the interaction between two proteins, LCA and LTF or STF, using fluorescein as a probe. Fluorescein is bound covalently to LCA, and so its fluorescence is sensitive to interactions occurring between the proteins. Since we are following binding experiments, we have static quenching inducing a decrease in fluorescence intensity of fluorescein (Figure 13.5). There is no need for LTF to STF bind to the fluorescein to observe a decrease in the fluorescence intensity of fluorescein. The external probe is part of LCA, and so its fluorescence is sensitive to structural modification(s) occurring in LCA.

The emission peak of fluorescein is still in the same position at all LTF concentrations. The reason for this is that the environment of the fluorescein on LCA was not affected or modified with binding of LTF on LCA. However, it is important to note that the fluorescence emission peak of fluorescein is not very sensitive to all modifications occurring in its proximity. Other probes have a much more sensitive emission peak than

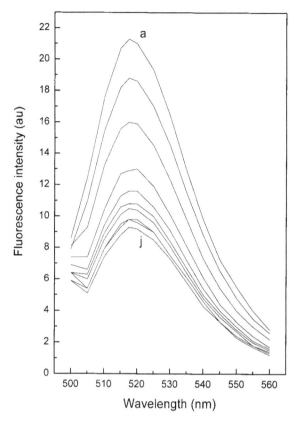

Figure 13.5 Fluorescence spectra of LCA–fluorescein in the absence (a) and presence of increasing concentrations of LTF. Spectrum j corresponds to 1.223 μM of LTF. λ_{ex} = 495 nm.

that of fluorescein, which is why it is very important not to generalize the fluorescence properties to all fluorophores.

b Why do we need to use an extrinsic probe to perform our experiments instead of adding LTF or STF to LCA and following the fluorescence intensity variation of Trp residues of LCA?

The three proteins show intrinsic fluorescence due to the presence of Trp residues. Thus, although binding occurs between LCA and LTF or STF, we cannot use fluorescence of Trp residues of LCA to follow this interaction, since there will be an overlapping with the fluorescence of Trp residues of LTF or STF. Therefore, it is necessary to use an extrinsic probe, which is bound to one protein only. A covalently bound fluorophore such as fluorescein is very suitable to perform binding experiments, since there will be no real binding between the fluorophore and the added protein.

c Figure 13.6 shows the fluorescence intensity at 515 nm of fluorescein bound to LCA in the presence of increasing concentrations of LTF or STF.

The plots clearly indicate that the affinity of LTF to LCA is much higher than the affinity of STF to LCA. In fact, the curvature observed for the LTF–STF interaction

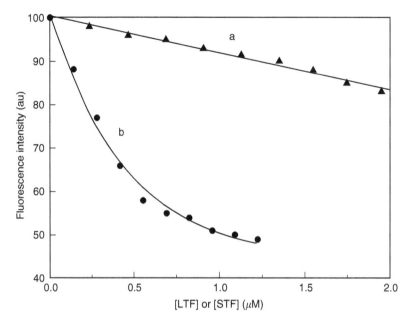

Figure 13.6 Fluorescence intensity quenching of fluorescein bound to LVA, as a result of STF–LCA (a) and of LCF–LCA (b) interactions. [LCA] = 0.7 μM. Source: Albani, J. R., Debray, H., Vincent, M. and Gallay, J. (1997). *Journal of Fluorescence*, **7**, 293–298. Albani, J.R., Debray, H., Vincent, M. and Gallay, J. (1997). *Journal of Fluorescence*, **7**, 293–298. Figure No. 1. With kind permission of Springer Science and Business Media (1, 2, and 3).

means that a complex is obtained. However, the absence of such curvature for the STF–LCA interaction means that we are far from reaching a complex between the two proteins.

d The value of ΔF_{max} for the LCA–LTF complex was obtained by plotting $1/\Delta F$ as a function of $1/[LTF]$. The value found is 69.2 (Figure 13.7). Since the data for STF binding on LCA do not show a curvature, it is not possible to calculate the ΔF_{max} of the STF–LCA interaction by plotting $1/\Delta F$ as a function of $1/[STF]$. In fact, plotting the inverse of the linear region in Figure 13.6. Figure 13.8 does not allow the determination of the ΔF_{max} calculated from the inverse of all the data (Figure 13.7). Therefore, we have to take 69.2 as the value of the maximum fluorescence change for both LCA–LTF and LCA–STF interactions.

From a hyperbolic modification of any spectral parameter as a function of the added ligand concentration, the reciprocal function of the hyperbolic can be plotted. In fact, a reciprocal function of a hyperbolic is linear, while a reciprocal function of a linear plot is not linear. Therefore, in order to determine the value of ΔF_{max} from the experiments of Figure 13.6b, it is important to plot the reciprocal function using all the data, mainly those in the curvature, since these data are part of the curvature and are the closest to the value of ΔF_{max} we are looking after.

e Equation (13.3) yields K_d values of 0.1035 and 5.3 μM for the LCA–LTF and LCA–STF complexes, respectively. The equivalent association constants are 9.66 and 0.188 μM^{-1}.

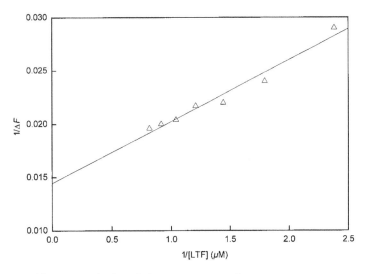

Figure 13.7 Double reciprocal plot of the LCA–FITC–LTF fluorescence intensity variation vs LTF concentration.

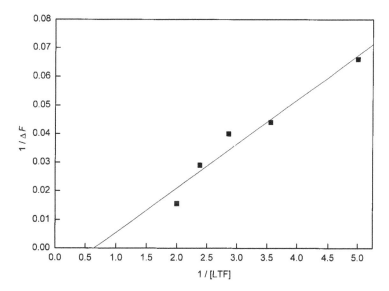

Figure 13.8 Double-reciprocal plot of the LCA–FITC–LTF fluorescence intensity variation vs LTF concentration. The data are obtained from the linear part (low concentrations of LTF) of the graph of Figure 13.6b.

Thus, LTF has its affinity 50 times higher to LCA than STF. The reason for this is the presence of α-1,6-fucose bound to the N-acetylglucosamine residue.

f Plotting I_0/I vs [STF] or [LTF] yields linear graphs with slopes equal to the asociation constants of the complexes (not shown). The values of K_d were found to be 0.875 and 10 μM for the LCA–LTF and LCA–STF complexes, respectively. These values show that the affinity of LCA to LTF is 11 times higher than that of LCA for STF, and not 50 times as we found with Equation (13.3). One reason for this difference is the fact that Equation (10.22) does not take into consideration the concentration of nonbound STF or LTF.

References

Albani, J.R., Debray, H., Vincent, M. and Gallay, J. (1997). Role of the carbohydrate moiety and of the alpha l-fucose in the stabilization and the dynamics of the *Lens culinaris* agglutinin–glycoprotein complex. A fluorescence study. *Journal of Fluorescence*, **7**, 293–298.

Bourne, Y., Van Tilbeurgh, H. and Cambillau, C. (1993). Protein–carbohydrate interactions. *Current Opinion in Structural Biology*, **3**, 681–686.

Kornfeld, K., Reitman, M.-L. and Kornfeld, R. (1981). The carbohydrate-binding specificity of pea and lentil lectins. Fucose is an important determinant. *Journal of Biological Chemistry*, **256**, 6633–6640.

Tuma, J.J. (1983). *Handbook of Physical Calculations*, McGraw-Hill, New York.

Chapter 14
Förster Energy Transfer

14.1 Principles and Applications

Förster energy transfer or energy transfer at a distance occurs between two molecules, a donor (the excited fluorophore) and an acceptor (a chromophore or a fluorophore) (Forster 1948). Energy is transferred by resonance, i.e., the electron of the excited molecule induces an oscillating electric field that excites acceptor electrons. The latters will reach an excited state. If the acceptor is fluorescent, its de-excitation will occur mainly by a photon emission. However, if it does not fluoresce, it will return to a fundamental state as a result of its interaction with the solvent. Energy-transfer efficiency depends on three parameters, distance R between the donor and acceptor, spectral overlap between the donor fluorescence spectrum and the acceptor absorption spectrum (Figure 14.1), and the orientation factor κ^2, which gives an indication of the relative alignment of the dipoles of the acceptor in the fundamental state and of the donor in the excited state.

The distance that separates the two molecules goes from 10 to 60–100 Å. Below 10 Å, electron transfer may occur between the two molecules, inducing an energy transfer from donor to acceptor. κ^2 values hold from 0 to 4. For aligned and parallel transition dipoles (maximal energy transfer) κ^2 is 4, and if the dipoles are oriented perpendicular to each other (very weak energy transfer), κ^2 is 0. When κ^2 is not known, its value is considered to be equal to 2/3. This value corresponds to a random relative orientation of the dipoles.

The energy-transfer mechanism can be described by the following:

$$D + h\nu_0 \rightarrow D^* \qquad \text{absorption}$$
$$D^* \rightarrow D + h\nu_1 \qquad \text{fluorescence}$$
$$D^* \rightarrow D \qquad \text{nonradiative de-excitation}$$
$$D^* + A \rightarrow D + A^* \quad \text{energy transfer}$$
$$A^* \rightarrow A \qquad \text{nonradiative de-excitation}$$
$$A^* \rightarrow A + h\nu_2 \qquad \text{induced fluorescence}$$

Figure 14.2 shows fluorescence intensity quenching of 4′,6-diamidino-2-phenylindole (DAPI) complexed to DNA in the presence of two concentrations of Acridine Orange. In fact, one can see that while the fluorescence intensity of DAPI decreases, that of Acridine

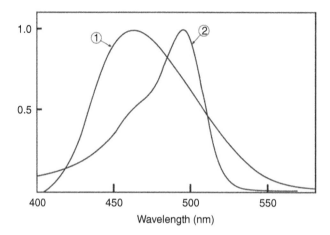

Figure 14.1 Normalized corrected emission spectrum of apoHb–ANS complex upon excitation at 350 nm. 2: Normalized absorption spectrum of apo FIA. Reprinted with permission from Sassaroli, M., Bucci, E., Liesegang, J., Fronticelli, C. and Steiner, R. F. (1984). *Biochemistry*, 2487–2491. Copyright © 1984 American Chemical Society.

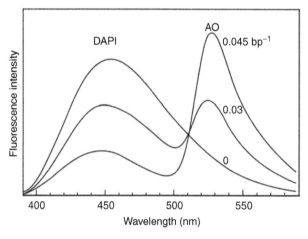

Figure 14.2 Fluorescence emission spectra of DNA-bound DAPI in the presence of 0.03 and 0.045 bp^{-1} of AO. The excitation wavelength is 360 nm. Source: Maliwal, B.P., Kusba, J. and Lakowicz, J.R. (1995). *Biopolymers*, **35**, 245–255. Reprinted with permission from John Wiley & Sons, Inc.

Orange increases. This type of experiment is one of several ways to show the presence of the energy-transfer mechanism.

Also, one can put into evidence the energy-transfer mechanism by recording the fluorescence excitation spectrum of the complex (donor–acceptor) (λ_{em} is set at a wavelength where only the acceptor emits) and comparing it to the absorption spectrum of the donor alone. In the presence of energy transfer between the two molecules, a peak characteristic of the donor absorption will be displayed in the fluorescence excitation spectrum.

Membranes fusion can be studied using the energy-transfer mechanism. In fact, membrane vesicles labeled with both NBD and rhodamine probes are fused with unlabeled vesicles. In the labeled vesicles, upon excitation of NBD at 470 nm, emission from rhodamine is observed at 585 nm as a result of energy transfer from NBD to rhodamine. The average distance separating the donor from the acceptor molecules increases with fusion of the vesicles, thereby decreasing the energy-transfer efficiency (Struck *et al.* 1981).

Energy transfer is considered as a type of dynamic quenching. In general, dynamic quenching can be described with three different processes: electron transfer, electron exchange, and Förster or coulombic energy transfer (Figure 14.3).

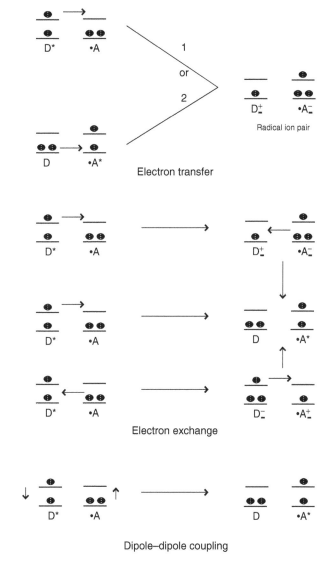

Figure 14.3 Different mechanisms of dynamic quenching.

Electron exchange and coulombic modes are part of what we can call in general the energy-transfer mechanism. In the coulombic mode, an electron transfer does not occur between donor and acceptor molecules, but two transitions are observed simultaneously. The rate of dipole–dipole transfer depends on the oscillator strength of the radiative transitions, which is not the case for the electron-exchange mechanism. Also, in the Förster mechanism, there is a dipole–dipole coupling mechanism, which is effective in the singlet–singlet energy transfer only, as a result of the large transition dipoles. Finally, both Förster and electron-exchange rate constants depend on the spectral overlap integral J between the emission spectrum of the donor and the absorption spectrum of the acceptor. A high value of J induces a high electron rate.

Combination of time-resolved fluorescence and Förster energy transfer was applied in the method known as HTRF® (homogeneous time-resolved fluorescence) developed by Cisbio (http://www.htrf-assays.com/). This allows the use of a lanthanide, such as europium, with an extremely long emission half-life (from μs to ms) and a large Stokes shift, compared to more traditional fluorophores. Conjugation of Eu^{3+} to cryptate gives an entity which confers increased assay stability and the use of a patented ratiometric measurement that allows correction for quenching and sample interferences. This powerful combination provides significant benefits to drug-discovery researchers including assay flexibility, reliability, increased assay sensitivity, higher throughput, and fewer false-positive/false-negative results.

Many compounds and proteins present in biological fluids or serum are naturally fluorescent, and the use of conventional, prompt fluorophores leads to serious limitations in assay sensitivity. The use of long-lived fluorophores combined with time-resolved detection (a delay between excitation and emission detection) minimizes any prompt fluorescence interference. However, it is difficult to generate fluorescence of lanthanide ions by direct excitation, because of the ions' poor ability to absorb light. Lanthanides must first be complexed with organic moieties that harvest light and transfer it to the lanthanide through intramolecular, nonradiative processes. Rare earth chelates and cryptates are examples of light-harvesting devices. The collected energy is transferred to the rare earth ion, which then emits its characteristic long-lived fluorescence.

To be successfully used as labels in biological assays, rare earth complexes should possess specific properties including stability, high light yield, and ability to be linked to biomolecules. Moreover, insensitivity to fluorescence quenching is of crucial importance when working directly in biological fluids. When complexed with cryptates, however, many of these limitations are eliminated.

In HTRF®, the donor is europium cryptate, which has the long-lived emissions of lanthanides coupled with the stability of cryptate encapsulation, and the primary acceptor fluorophore is XL665, a modified allophycocyanin. When these two fluorophores are brought together by a biomolecular interaction and after excitation, some of the energy captured by the Cryptate is released through fluorescence emission with a peak at 620 nm, while the remaining energy is transferred to XL665. This energy is then released by XL665 as specific fluorescence at 665 nm. Emission at 665 nm occurs only through FRET with europium. Because europium cryptate is always present in the assay, light at 620 nm is detected even when the biomolecular interaction does not bring XL665 within close proximity.

HTRF® is a highly flexible chemistry and has been successfully used to measure molecular complexes of many different sizes. This includes assessment of small phosphorylated peptides, immunoassays for quantifying large glycoproteins such as thyroglobulin, receptor tyrosine kinase activity using membrane preparations, and indirect detection (via secondary antibodies) of tagged complexes such as CD28/CD86 binding.

Heparanase, a heparan sulfate-specific endo-β-d-glucuronidase, cleaves heparin sulfate (HS) into characteristic large-molecular-weight fragments. HS degradation causes the release of angiogenic and growth factors which lead to tumor-cell proliferation and migration, and to angiogenesis. Heparanase is therefore an attractive target for the development of new anti-metastatic therapeutics. Chikachi *et al.* developed a highly sensitive, quantitative, and robust heparanase assay using HTRF® technology and based on time-resolved fluorescence energy transfer (TR-FRET) between europium cryptate and XL665 (allophycocyanin) (Figure 14.4).

Cryptates are formed by the inclusion of a cation into a tridimensional cavity. The properties of the trisbypyridine macrocycle, for instance (Figure 14.5), favor such a tight

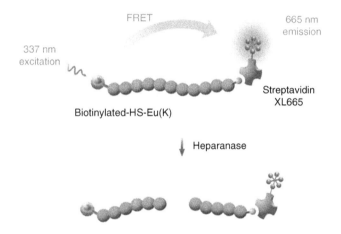

Figure 14.4 Heparanase assay configuration. Courtesy of Cisbio.

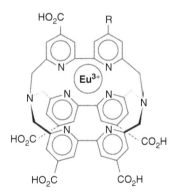

Figure 14.5 Structure of trisbypirydine pentacarboxylate Eu^{3+} cryptate. Courtesy of Cisbio.

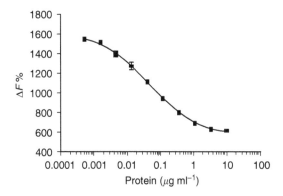

Figure 14.6   Heparanase activity titration curve.

association of the cage with the europium in that this interaction becomes virtually unchallengeable, and leads to an exceptionally inert complex.

The classical TBP Eu^{3+} has its excitation peak at 307 nm and, substituting carboxylic groups on two of the bipyridine units, leads to a shift in the excitation peak to 325 nm. Thus, excitation with a nitrogen laser at 337 nm induces an increase in the cryptate fluorescence by 2.5. Energy transfer was observed when an HS substrate labeled with cryptate and biotin were incubated with streptavidin-XL665. In this assay configuration, the maximum signal was obtained without enzymes. Cleavage of the biotin-HS-cryptate substrate by heparanase prevented FRET from occurring and therefore led to a signal decrease (Figure 14.6).

14.2   Energy-transfer Parameters

Let us consider k_t as the rate constant of the excited state depopulation via the energy-transfer mechanism; the measured fluorescence lifetime is thus equal to

$$\frac{1}{\tau_t} = \frac{1}{\tau_o} + k_t \tag{14.1}$$

where $1/\tau$ is the de-excitation rate constant in the absence of energy transfer. The efficiency E of de-excitation by means of the energy-transfer mechanism is

$$E = \frac{k_t}{k_t + 1/\tau_o} = \frac{1/\tau_t - 1/\tau_o}{1/\tau_t} = 1 - \frac{1/\tau_o}{1/\tau_t} = 1 - \frac{\tau_t}{\tau_o} \tag{14.2}$$

The energy-transfer mechanism appears to affect the fluorescence lifetime, intensity, and quantum yield:

$$E = 1 - \frac{I}{I_o} = 1 - \frac{\phi_t}{\phi_o} = 1 - \frac{\tau_t}{\tau_o} \tag{14.3}$$

where τ, I, and ϕ_t are the mean fluorescence lifetime and intensity in the absence (τ_o, I_o, and ϕ_o) and presence of the quencher (τ_t, I, and ϕ_t).

The distance R that separates the donor from the acceptor is calculated using Equation (14.4):

$$R = R_0 \left(\frac{1 - E}{E} \right)^{1/6} \tag{14.4}$$

The Förster distance $R_o(\text{Å})$ at which the energy-transfer efficiency is 50% is calculated using Equation (14.5):

$$R_0 = 9.78 \times 10^3 [\kappa^2 n^{-4} Q_D J(\lambda)]^{1/6} \tag{14.5}$$

where κ^2 is the orientation factor ($=2/3$), n the refractive index ($=1.33$) and Q_D the average quantum yield of the donor in the absence of the acceptor.

From the overlap of the emissium spectrum of the donor and the absorption spectrum of the acceptor, we can calculate the overlap integral J ($\text{M}^{-1}\text{cm}^{3)}$)

$$J(\lambda) = \frac{\int_0^\infty F_D(\lambda) \cdot \varepsilon_A(\lambda) \cdot \lambda^4 \, d\lambda}{\int_0^\infty F_D(\lambda) \, d\lambda} \tag{14.6}$$

ε_A is the extinction coefficient of the acceptor expressed in $\text{M}^{-1}\text{cm}^{-1}$, F_D the fluorescence intensity of the donor expressed in arbitrary units, and λ the wavelength (cm).

The distance R that separates the donor from the acceptor can be determined from X-ray diffraction studies. The κ^2 value of the five tryptophan residues of cytochrome P-450 has been determined. Cytochrome P-450, a superfamily of heme proteins, catalyses monooxygenation of a wide variety of hydrophobic substances of xenobiotic and exogenous origin (Guengerich 1991; Porter and Coon 1991). Cytochrome P-450 from *Bacillus megaterium* strain ATCC 14581 called cytochrome P-450$_{BM3}$ is a monomer of molecular mass of 119 kDa, comprising an N-terminal P-450 heme domain linked to a C-terminal reductase domain (Narhi and Fulco 1986). The heme domain of cytochrome P-450$_{BM3}$ contains five tryptophan residues (Figure 14.7). The distances of the tryptophans from

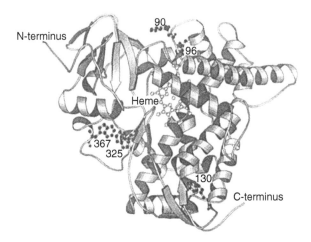

Figure 14.7 Location of five tryptophan residues (shown in a ball-and-stick representation) and the heme prosthetic group (white ball and stick) in the heme domain of cytochrome P-450$_{BM3}$.

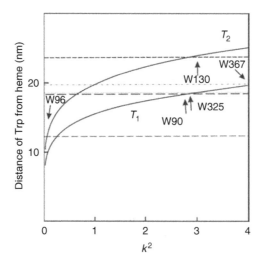

Figure 14.8   Plot of κ^2 against the distance of tryptophan residues from the heme in cytochrome P-450. The dotted lines in the figure show the crystallographic distances between tryptophan residues, and lines labeled with τ_x and τ_2 show the calculated distances between the heme and different tryptophan residues using all possible values of κ^2 (0.0–4.0). The point where these two lines intersect shows the κ^2 value for the particular residue and is indicated by the arrows. Source of figures 11.22 and 11.23: Khan, K.K., Mazumdar, S., Modi, S., Sutcliffe, M., Roberts, G.C.K. and Mitra, S. (1997). *European Journal of Biochemistry*, 244, 361–370.

the heme are 1.224 (Trp96), 1.834 (Trp90), 1.85 (Trp325), 1.968 (Trp367), and 2.365 nm (Trp130). Fluorescence intensity decay yields three fluorescence lifetimes equal to 0.2, 1, and 5.4 ns with fractional contributions of 61%, 33%, and 6%, respectively.

Figure 14.8 shows the plot of Trp to heme distance obtained by fluorescence energy transfer as a function of κ^2 (from 0 to 4) for the two shortest fluorescence lifetimes (solid curve). The five dotted lines in the figure show the crystallographic distances between tryptophan residues and heme. The intersection between the crystallographic lines and the energy-transfer plots is equal to the κ^2 corresponding to the specific tryptophan.

14.3   Bioluminescence Resonance Energy Transfer

Bioluminescence resonance energy transfer (BRET) takes advantage of the Förster resonance energy transfer and is observed in the sea pansy *Renilla reniformis*. This organism expresses a luciferase, which emits blue light when it is purified. If the luciferase is excited in intact cells, green light occurs, because *in vivo* the luciferase is associated with the green fluorescent protein (GFP), which accepts the energy from the luciferase and emits green light. BRET occurs in the 1–10 nm regions, which is comparable with the dimensions of biological macromolecules and makes it an ideal system for the study of protein–protein interaction in living cells (Ayoub *et al.* 2002).

Green fluorescent protein is commonly used for energy-transfer experiments (Baubet *et al.* 2000). The fluorescent moiety of GFP protein is the Ser–Tyr–Gly derived chromophore. GFP can be expressed in a variety of cells where it becomes fluorescent, can be fused to a host protein, and can be mutated so that the mutants have different fluorescence properties and can be used in energy-transfer studies. Some examples, the blue fluorescent protein (BFP), which is a GFP mutant with a Tyr66His mutation, absorbs at 383 nm and emits at 447 nm. In energy-transfer experiments, BFP is a donor molecule to GFP. In a system with randomly oriented chromophores, the distance R_0 at which 50% energy transfer occurs is found to be equal to 4 nm (Heim 1999). Cyan fluorescent protein (CFP) contains a Tyr66Trp mutation and absorbs and emits at 436 and 476 nm, respectively. Yellow fluorescent protein (YFP) is a Thr203Tyr mutant with excitation and emission peaks equal to 516 and 529 nm, respectively. In the CFP–YFP pair, CFP is the donor molecule, and the YFP is the acceptor molecule. The value of R_0 found for randomly oriented chromophores is 5.2 nm. The CFP–YFP pair fused to proteins has been used in resonance energy-transfer studies along with multifocal multiphoton microscopy to measure transport phenomena in living cells (Majoul *et al.* 2002) and protein–protein interaction and structural changes within a molecule (Truong and Ikura 2001).

Also, the Phe-64 $\rightarrow$ Leu and Ser-65 $\rightarrow$ Thr mutant shows higher fluorescence parameters than that of the wild GFP. Excitation at 458 nm yields a fluorescence spectrum with two peaks at 512 and 530 nm. The fluorescence properties of this enhanced green fluorescent protein (EGFP) were found to be similar to the recombinant glutathione *S*-transferase–EGFP (GST–EGFP) protein, expressed in *Escherichia coli* (Cinelli *et al.* 2004).

Imaging fluorescence resonance energy-transfer studies between two GFP mutants in living yeast have also been performed. The donor was the GFP mutant P4.3, bearing the Y66H and Y145F mutations, and the acceptor was the GFP mutant S65T (Sagot *et al.* 1999). The authors constructed a concatemer where the two GFP mutants were linked by a spacer containing a protease-specific recognition site. The tobacco etch virus (TEV) protease was added to cleave the covalent bond between the two fluorophores. Experiments performed *in vitro* show that excitation of the donor at 385 nm leads to the emission of both the donor and the acceptor at emission peaks equal to 385 and 445 nm, respectively. Addition of TEV protease induces an increase in the donor emission and a decrease of the emission acceptor. Experiments performed *in vivo* reveal that in the presence of protease, the energy-transfer loss is around 57%.

In most BRET applications, the fused donor is *Renilla*luciferase (Rluc) rather than aequorin, to avoid any intrinsic affinity for *Aequorea*-derived GFP mutant; the acceptor is the YFP, to increase the spectral distinction between the two emissions. When the donor and acceptor are in close proximity, the energy resulting from catalytic degradation of the coelenterazine derivative substrate is transferred from the luciferase to the YFP, which will then emit fluorescence at its characteristic wavelength (Xu *et al.* 1999).

To demonstrate the clear discrimination between positive and negative control of the BRET assay technology, the luminescence and fluorescence signals of the BRET[2™] demo kit (Perkin Elmer Life Sciences) were quantified on the microplate reader POLARstar OPTIMA (BMG Labtech: http://www.pharmaceutical-technology.com/contractors/lab_equip/bmg_labtech/), allowing the monitoring of the kinetic curves and the calculation of the BRET ratio. The BRET[2™] demo kit applies the cell-permeable and nontoxic coelenterazine derivative substrate DeepBlueC™ (DBC) and a mutant of the GFP[2] as acceptor. These compounds show improved spectral resolution and sensitivity over earlier variants. Four sets of samples were run in triplicate, a blank (non-transfected cells), a positive control (Rluc-GFP[2]), a negative control (Rluc + GFP[2]), and a buffer control

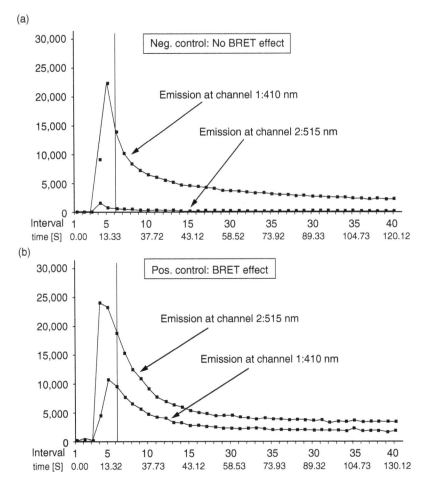

Figure 14.9 (a), (b) Discrimination between negative and positive control of the BRET assay technology. Resonance energy transfer is obvious for the positive control. No BRET occurs for the negative control. Courtesy of BMG Labtech, Germany.

($BRET^2$ assay buffer). Readings were started immediately after the automated injection of the luciferase substrate DBC. The kinetic curves of the negative control are shown in Figure 14.9a, for both channels. The low values of the 515 nm channel indicate that no resonance energy transfer occurred. Whereas the positive control shows reduced values at the 410 nm and elevated values at the 515 nm channel due to the BRET effect (Figure 14.9b). In both figures, the fluorescence intensities at the two channels were not recorded simultaneously. The low values of the 515 nm channel indicate that no resonance energy transfer occurred, whereas the positive control shows reduced values at the 410 nm and elevated values at the 515 nm channel due to the BRET effect.

Artificial fusion construct of the positive control (Rluc-GFP^2) induces an extremely high BRET. Also, the large spectral resolution between donor and emission peaks in BRET (115 nm) (Angers *et al.* 2000) greatly improves the signal to background ratio over

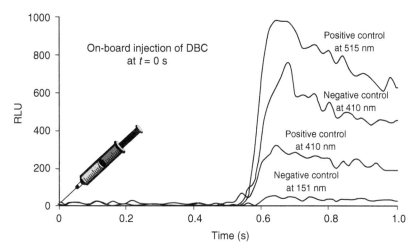

Figure 14.10 Kinetics curves observed simultaneously in two channels with and without energy transfer. Courtesy of BMG Labtech, Germany.

traditionally used BRET and FRET technologies that typically have only a ~50 nm spectral resolution (Mahajan *et al.* 1998).

In another experiment, the two channels were recorded simultaneously, and kinetic curves (raw data–blank) of the negative control are shown in Figure 14.10 for both channels.

The calculated BRET ratio indicates the occurrence of protein–protein interaction *in vivo*. In Figure 14.11, the blank corrected BRET$^{2^{TM}}$ ratios for both negative and positive controls are shown and were determined as

$$\text{BRET}^{2^{TM}} \text{ ratio} = \frac{\text{emission at 515 nm} - \text{emission at 515 nm of nontransfected cells}}{\text{emission at 410 nm} - \text{emission at 410 nm of nontransfected cells}}$$

$$(14.7)$$

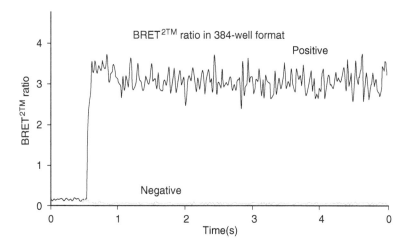

Figure 14.11 Ratio of negative and positive control over time. Courtesy of BMG Labtech, Germany.

The signal for negative and positive control here reveals a value of around 0.06 and 3.3, respectively, which leads to a factor of around 50 and a clear discrimination between these controls. The high factor between these controls is caused by the artificial fusion construct of the positive control (Rluc-GFP^2) resulting in an extremely high BRET.

The main disadvantages of FRET, as opposed to BRET, are the consequences of the required excitation of the donor with an external light source. BRET assays show no photo-bleaching or photo-isomerization of the donor protein, no photo-damage to cells, and no light scattering or autofluorescence from cells or microplates, which can be caused by incident excitation light. In addition, one main advantage of BRET over FRET is the lack of emission arising from direct excitation of the acceptor. This reduction in background should permit detection of interacting proteins at much lower concentrations than is possible for FRET. However, BRET requires the addition of a cofactor, and for some applications, e.g., determining the compartmentalization and functional organization of living cells, the GFP-based FRET method is superior to BRET due to the much higher light output.

In conclusion, BRET offers the ability to directly study complex protein–protein interactions in living cells. There is no need for an excitation light source. Therefore, photosensitive tissue can be used for BRET, and problems associated with FRET-based assays such as photo-bleaching, autofluorescence, and direct excitation of the acceptor are eliminated.

References

Angers, S., Salahpour, A., Joly, E., Hilairet, S., Chelsky, D., Dennis, M. and Bouvier, M. (2000). Detection of b$_2$-adrenergic receptors dimerization in living cells using bioluminescence resonance energy transfer (BRET). *Proceedings of the National Academy of Sciences, USA*, **97**, 3684–3689.

Ayoub, M.A., Couturier, C., Lucas-Meunier, E., Angers, S., Fossier, P., Bouvier, M. and Jockers, R. (2002). Monitoring of ligand-independent dimerization and ligand-induced conformational changes of melatonin receptors in living cells by bioluminescence resonance energy transfer. *Journal of Biological Chemistry*, **277**, 21522–21528.

Baubet, V., Le Mouellic, H., Campbell, A.K., Lucas-Meunier, E., Fossier, P. and Brûlet, P. (2000). Chimeric green fluorescent protein-aequorin as bioluminescent Ca^{2+} reporters at the single-cell level. *Proceedings of the National Academy of Sciences, USA*, **97**, 7260–7265.

Cinelli, R.A.G., Ferrari, A., Pellegrini, V., Tyagi, M., Giacca, M. and Beltram, F. (2004). The enhanced green fluorescent protein as a tool for the analysis of protein dynamics and localization: Local fluorescence study at the single-molecule level. *Photochemistry and Photobiology*, **71**, 771–776.

Forster, T. (1948). Intermolecular energy migration and fluorescence. *Annual of Physics (Leipzig)*, **2**, 55–75.

Guengerich, F.P. (1991). Reactions and significance of cytochrome P-450 enzymes. *Journal of Biological Chemistry*, **266**, 10019–10022.

Heim, R. (1999). Green fluorescent protein forms for energy transfer. *Methods in Enzymology*, **302**, 408–423.

Khan, M.J., Joginadha Swamy, M., Krishna Sastry, M.V., Umadevi Sajjan, S., Patanjali, S.R., Rao, P., Swarnalatha, G.V., Banerjee, P. and Surolia, A. (1988). Saccharide binding to three Gal/GalNac specific lectins: Fluorescence, spectroscopic and stopped-flow kinetic studies. *Glycoconjugate Journal*, **5**, 75–84.

Khan, K.K., Mazumdar, S., Modi, S., Sutcliffe, M., Roberts, G.C.K. and Mitra, S. (1997). Steady-state and picosecond-time-resolved fluorescence studies on the recombinant heme domain of *Bacillus megaterium* cytochrome P-450. *European Journal of Biochemistry*, **244**, 361–370.

Mahajan, N.P., Linder, K., Berry, G., Gordon, G.W., Heim, R. and Herman, B. (1998). Bcl-2 and Bax interactions in mitochondria probed with green fluorescent protein and fluorescence resonance energy transfer. *Nature Biotechnology*, **16**, 547–552.

Maliwal, B.P., Kusba, J. and Lakowicz, J. R. (1995). Fluorescence energy transfer in one dimension: Frequency-domain fluorescence study of DNA–fluorophore complex. *Biopolymers*, **35**, 245–255.

Majoul, I., Straub, M., Duden, R., Hell, S.W., Soling, H.D. (2002). Fluorescence resonance energy transfer analysis of protein–protein interactions in single living cells by multifocal multiphoton microscopy. *Journal of Biotechnology*, **82**, 267–277.

Narhi, L.O. and Fulco, A.J. (1986). Characterization of a catalytically self-sufficient 119,000-dalton cytochrome P-450 monooxygenase induced by barbiturates in *Bacillus megaterium*. *Journal of Biological Chemistry*, **261**, 7160–7169.

Porter, T.D. and Coon, M.J. (1991). Cytochrome P-450. Multiplicity of isoforms, substrates, and catalytic and regulatory mechanisms. *Journal of Biological Chemistry*, **266**, 13469–13472.

Sagot, I., Bonneu, M., Balguerie, A. and Aigle, M. (1999). Imaging fluorescence resonance energy transfer between two green fluorescent proteins in living yeast. *FEBS Letters*, **447**, 53–57.

Sassaroli, M., Bucci, E., Liesegang, J., Fronticelli, C. and Steiner, R.F. (1984). Specialized functional domains in hemoglobin: dimensions in solution of the apohemoglobin dimer labeled with fluorescein iodoacetamide. *Biochemistry*, **23**, 2487–2491.

Struck, D.K., Hoekstra, D. and Pagano, R.E. (1981). Use of resonance energy transfer to monitor membrane fusion. *Biochemistry*, **20**, 4093–4099.

Truong, K. and Ikura, M. (2001). The use of FRET imaging microscopy to detect protein–protein interactions and protein conformational changes *in vivo*. *Current Opinion in Structural Biology*, **11**, 573–578.

Xu, Y., Piston, D.W. and Johnson, C.H. (1999). A bioluminescence resonance energy transfer (BRET) system: application to interacting circadian clock proteins. *Proceedings of the National Academy of Sciences, USA*, **96**, 151–156.

Chapter 15
Binding of TNS on Bovine Serum Albumin at pH 3 and pH 7

15.1 Objectives

The purpose of this experiment is to find out how pH affects proteins tertiary structure and thus the binding of a ligand to proteins. In principle, when a protein is dissolved in a buffer with a pH far from its physiological pH, its tertiary structure is altered. Therefore, we can speak of partial or complete denaturation or unfolding of the protein. The characteristics of an unfolded protein such as ligand binding differ from those of the folded protein. One can follow the protein–ligand interaction by recording the fluorescence emission parameters (intensity, anisotropy, and lifetime) of the protein Trp residues and of the ligand if it is fluorescing. In the present work, students will study the structural alteration of bovine serum albumin (BSA) with pH, by following the fluorescence of both protein Trp residues and of TNS.

15.2 Experiments

15.2.1 Fluorescence emission spectra of TNS–BSA at pH 3 and 7

Plot the fluorescence emission spectra from 290 to 500 nm ($\lambda_{ex} = 280$ nm) of 1.5 μM BSA solutions prepared at pH 3 and 7. Then, to each cuvette, add 15 μM TNS and record the protein emission spectra at the two pHs. Correct the fluorescence intensity for the inner filter effect, and then plot the corrected emission spectra. What do you notice? Are the spectra obtained at the two pHs identical? What are the values of the emission peaks you observe? Explain the results you obtained.

Do not forget to measure the optical densities at the excitation wavelength and along the emission spectra ε of bovine serum albumin at 280 nm is 43 824 M^{-1} cm^{-1} and of TNS at 317 nm, 1.89×10^4 M^{-1} cm^{-1}.

15.2.2 Titration of BSA with TNS at pH 3 and 7

To a cuvette containing 1.5–2 μM bovine serum albumin solution, add equal volumes of TNS solution so that you have 0.3–15 μM TNS in the cuvette. For each TNS concentration

(0–15 μM), measure the optical densities at 280, 317, and 335 nm, and plot the emission spectrum from 290 to 500 nm ($\lambda_{ex} = 280$ nm). The experiment should be performed at pHs 3 and 7. What can you say about the variation of the fluorescence emission intensities? Explain the phenomena you are observing. Do you observe an isoemissive point? Explain.

Measure the fluorescence intensities at 335 nm at the different TNS concentrations, and correct them first for the dilution then for the inner filter effect. Finally, plot the fluorescence intensity at 335 nm as a function of TNS concentration. What do you observe?

For each TNS concentration, measure the intensity variation ΔI, and then plot $1/\Delta I$ as a function of $1/[\text{TNS}]$. By extrapolating to $1/[\text{TNS}] = 0$, we can calculate ΔI_{max}. Is ΔI_{max} the same for both pHs? Explain. Determine the apparent value of the association constant K_a of the TNS–BSA complex. Are the association constants equal? Explain.

Plot the ratio I_0/I vs. [TNS] for each pH and calculate the apparent value of the association constant K_a of TNS–BSA complex. Are the association constants calculated identical to those obtained by plotting the inverse equation $1/\Delta I$ vs. $1/[\text{TNS}]$?

15.2.3 Measurement of energy-transfer efficiency from Trp residues to TNS

From the data obtained in the titration experiments, calculate the energy-transfer efficiency, E

$$E = 1 - \frac{I}{I_0} \tag{14.3}$$

for each TNS concentration, then determine the value of E at infinite concentrations of TNS by plotting $1/E$ vs. $1/[\text{TNS}]$. Is the value of E the same for both pHs? Explain.

15.2.4 Interaction between free Trp in solution and TNS

Repeat the experiment between TNS and BSA by replacing BSA with free L-Trp. What do you observe? By plotting I_0/I vs. [TNS] what is the nature of the constant you will calculate?

15.3 Results

Figure 15.1 shows the fluorescence emission spectrum of BSA Trp residues in the absence (spectrum a) and presence of TNS (b), both recorded at pH 3. First, we see that the fluorescence emission peak (325 nm) of the Trp residues of the protein decreases to almost zero in the presence of TNS, while an important emission peak at 435 nm characterizing the fluorescence emission of TNS bound to BSA appears.

The decrease in fluorescence intensity of the Trp residues is the result of energy transfer that is occurring from aromatic residues to bound TNS. The fluorescence of the ligand increases as a result of its binding to BSA and of the energy transfer that is occurring from Trp residues to TNS.

Figure 15.2 shows fluorescence emission spectra of BSA in the absence (a) and presence (b) of TNS, both recorded at pH 7. The fluorescence emission peak of Trp residues is

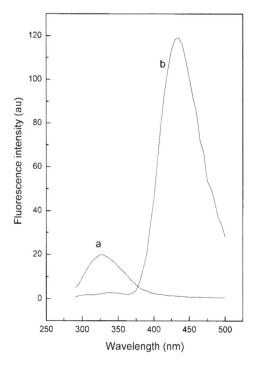

Figure 15.1   Fluorescence emission of BSA recorded in the absence (a) and presence of TNS (b) at pH 3. Spectra are corrected for the inner filter effect. $\lambda_{ex} = 280$ nm.

located at 335 nm and is still observed in the presence of TNS. This means that the energy transfer between Trp residues and TNS is less important at pH 7 than at pH 3 and/or the interaction between TNS and BSA is weaker at pH 7 than at pH 3. We can observe also that the decrease in fluorescence intensity of the Trp residues is associated with the increase in the fluorescence intensity of bound TNS. In fact, for the same concentrations of BSA ($2 \mu M$) and of TNS ($15 \mu M$), the fluorescence intensity increase in bound TNS is less important at pH 7 than at pH 3. Thus, one should conclude that the affinity of TNS to the protein at pH 3 is higher than its affinity at pH 7 inducing a better interaction between TNS and BSA at pH 3. Also, one should not rule out the possibility of having more binding sites for TNS on BSA at pH 3 than at pH 7.

Titration of a constant concentration of BSA with TNS yields a decrease in the fluorescence intensity of the Trp residues together with an increase in the fluorescence intensity of TNS (not shown). We do not observe an isoemissive point at either pH, thus indicating that TNS has more than one binding site on BSA. After correcting the fluorescence intensities at the emission peak of Trp residues for the dilution and then for the optical densities at the excitation ($\lambda_{ex} = 280$ nm) and emission wavelengths (335 or 325 nm), we can plot the fluorescence intensity as a function of TNS concentration at pH 3 and 7 (Figure 15.3). Results indicate that the fluorescence-intensity decrease is more important at pH 3 than at pH 7. This means that the interaction between TNS and BSA is more important at a low pH.

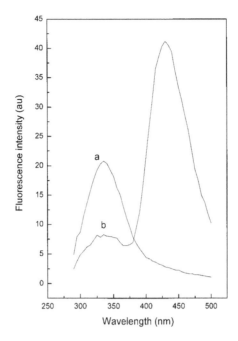

Figure 15.2 Fluorescence emission of BSA obtained in the absence (a) and presence of TNS (b) at pH 7. The spectra are corrected for the inner filter effect. $\lambda_{ex} = 280$ nm.

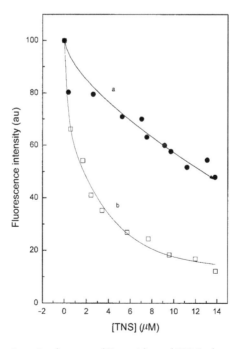

Figure 15.3 Fluorescence-intensity decrease of Trp residues of BSA in the presence of TNS at pH 7 (a) and pH 3 (b).

Plotting the variation ΔI of fluorescence decrease as a function of added TNS concentration yields a hyperbolic plot at both pHs (Figure 15.4). However, this hyperbolic plot is more obvious at pH 3 than at pH 7. Hyperbolic plots can be described by the following equation:

$$\Delta I = \frac{\Delta I_{max} D}{\Delta I_{max} + K_d} \tag{15.1}$$

where D is the concentration of added TNS, K_d the dissociation constant of TNS–BSA complex, and ΔI_{max} the maximal variation when TNS binding site on BSA is saturated.

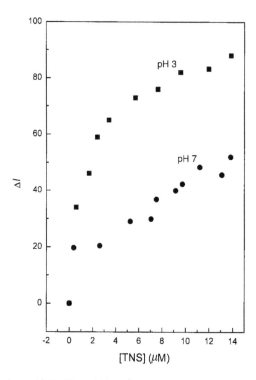

Figure 15.4 Variation of ΔI with [TNS] at pH 3 and 7.

The inverse plot of a hyperbolic equation is linear, and so the inverse plot of Equation (15.1) is

$$\frac{1}{\Delta I} = \frac{1}{\Delta I_{max}} + \frac{K_d}{\Delta I_{max} D} \tag{15.2}$$

Plotting $1/\Delta I$ vs. $1/[TNS]$ yields a straight line with a slope equal to $K_d/\Delta I_{max}$ and intercept equal to $1/\Delta I_{max}$.

Figure 15.5 shows inverse plots obtained at pH 7 and 3. The plots are obtained by using all the measurements performed. This yields a value of ΔI_{max} of 96.52 and 67.52 at pH 3 and 7, respectively, with dissociation constants equal to 1.74 and 6.28 μM, respectively.

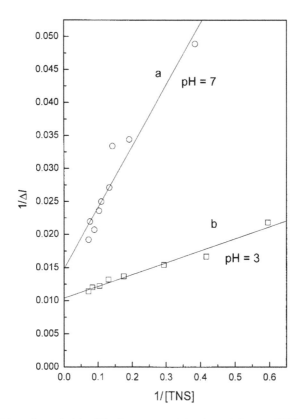

Figure 15.5 Double reciprocal plots of the fluorescence-intensity variation vs. [TNS] concentration.

The value of ΔI_{max} obtained at pH 7 is lower than the value obtained at the highest concentration of TNS reached in the experiment. The reason for this is the choice of the data used to plot Figure 15.5. In fact, we notice that the lowest values of TNS have been taken into consideration to plot the graph, while the highest values were not. This is a mistake, since the maximum value obtained at saturation is closer to the highest measured value than the lowest. Also, there is no reason to have a maximum value at pH 7 at saturation lower than that obtained at pH 3. Therefore, plotting the graph at pH 7 by considering the highest eight values of TNS and ΔI will yield a value for ΔI_{max} very similar to that obtained at pH 3 (Figure 15.6). The dissociation constant of TNS–BSA complex calculated at pH 7 is found to be 13 μM. Thus, the affinity of TNS to BSA is around seven times higher at pH 3 than at pH 7.

Binding of TNS to BSA can also be analyzed by the following equation:

$$\frac{I_0}{I} = 1 + K_a[\text{TNS}] \tag{15.3}$$

where K_a is the association constant of the TNS–BSA complex, and I_0 and I are the fluorescence intensities in the absence and presence of TNS. Plotting the data of Figure 15.3 with Equation (15.3), i.e., I_0/I vs. [TNS], yields the lines shown in Figure 15.7. The results

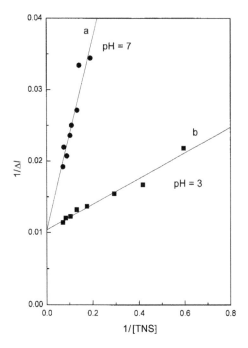

Figure 15.6 Determination of the maximal intensity variation at pH 3 and 7 using the highest values of ΔI and of [TNS].

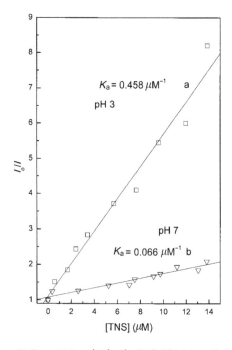

Figure 15.7 Plot of I_0/I vs. [TNS] at pH 3 and 7 for the TNS–BSA interaction.

indicate that the association constant measured at pH 3 is seven times higher than that measured at pH 7. This result is in good agreement with that obtained when calculations were performed using Equation (15.2) (Figure 15.6).

Application of Equations (15.2) and (15.3) in the determination of the dissociation constant of a complex implies that at every concentration of ligand we add, most of the ligand is bound to the protein, and so there is a small or negligible concentration of free ligand. When this is not the case, the application of Equations (15.2) and (15.3) is not appropriate, and another equation should be derived and applied to take into consideration the concentration of free ligand. Let us develop this equation. Binding of a ligand L to a protein P can be written as follows:

$$L + P \rightleftarrows LP \tag{15.4}$$

The dissociation constant of the complex is

$$K_d = \frac{[L]_f[P]_f}{[LP]} \tag{15.5}$$

where $[L]_f$ and $[P]_f$ characterize, respectively, the concentrations of free ligand and protein in solution.

The total concentration of ligand $[L]_o$ is

$$[L]_o = [L]_f + [LP] \tag{15.6}$$

The total protein concentration $[P]_o$ is

$$[P]_o = [P]_f + [LP] \tag{15.7}$$

Replacing Equations (15.6) and (15.7) in Equation (15.5) yields

$$K_d = \frac{([L]_o - [LP]) = ([P]_o - [LP])}{[LP]} \tag{15.8}$$

$$K_d = \frac{[L]_o[P]_o - [L]_o[LP] - [P]_o[LP] + [LP]^2}{[LP]} \tag{15.9}$$

$$K_d[LP] = [L]_o[P]_o - [L]_o[LP] - [P]_o[LP] + [LP]^2 \tag{15.10}$$

$$[LP]^2 - ([L]_o + [P]_o + K_d)[LP] + [L]_o[P]_o = 0 \tag{15.11}$$

We have an equation of second degree:

$$\Delta = ([P]_o + [L]_o + K_d)^2 - 4[P]_o[L]_o \tag{15.12}$$

$$[LP] = 0.5\{([L]_o + [P]_o + K_d) + \{([L]_o + [P]_o + K_d)^2 - 4[L]_o[P]_o\}^{1/2}\} \tag{15.13}$$

Knowing that

$$\Delta I = I_o = (\Delta I_{max}/I_o)([LP]/[P_o]) \tag{15.14}$$

where ΔI, ΔI_{max}, I_o, and $[P]$ are, respectively, the fluorescence-intensity variation of tryptophans at a concentration L_o of TNS, the fluorescence maximum variation when the binding site of TNS on the protein is saturated, the fluorescence intensity of the tryptophans in the absence of TNS, and the protein concentration.

Replacing Equation (15.14) in Equation (15.13) yields

$$\Delta I/I_o = (\Delta I_{max}/I_o)\times[([L]_o+[P]_o+K_d)+\{([L]_o+[P]_o+K_d)^2-4[L]_o[P]_o\}^{1/2}]/2[P]_o$$

$$(15.15)$$

The dissociation constant of the TNS–BSA complex at pH 3 is found to be equal to 1.1 μM, i.e., an association constant equal to 0.91 μM^{-1}. This value is in the same range as that obtained with Equations (15.2) and (15.3).

The energy-transfer efficiency E between Trp residues of BSA and TNS is calculated by plotting $1/E$ vs. $1/[TNS]$. Figure 15.8 indicates that the value of E is 94–96% for pH 3 and 7. Thus, at a low pH, BSA retains a specific conformation allowing adequate orientations of the two fluorophores TNS and Trp residues.

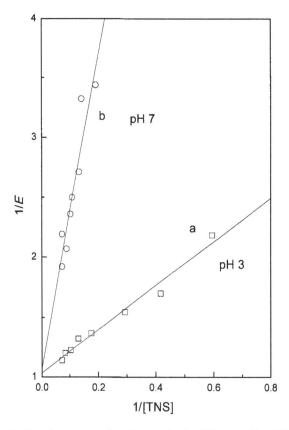

Figure 15.8 Determination of energy-transfer efficiency for the BSA Trp residues–TNS.

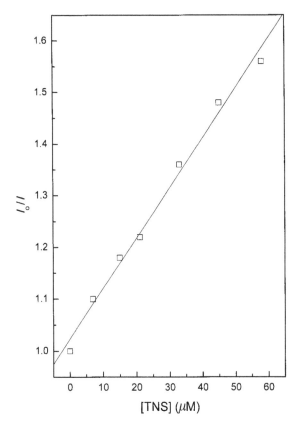

Figure 15.9 Stern–Volmer plot of fluorescence-intensity quenching of free Trp in solution with TNS. $K_{SV} = 0.0097 \ \mu M^{-1} = 9700 \ M^{-1}$. $k_q = K_{SV}/\tau_0 = 9700 \ M^{-1}/2.5 \ ns = 3.88 \times 10^{12} \ M^{-1} \ s^{-1}$.

Figure 15.9 shows a Stern–Volmer plot of fluorescence-intensity quenching with TNS of free L-Trp in solution. From the slope of the plot, a k_q value of around $4 \times 10^{12} \ M^{-1} \ s^{-1}$ can be obtained. This value is 100 times higher than that of bimolecular diffusion quenching of oxygen in solution. This means that fluorescence quenching by TNS of free tryptophan in solution is not controlled by diffusion but is the result of the Förster energy transfer (energy transfer at distance) occurring between the two fluorophores.

Chapter 16

Comet Test for Environmental Genotoxicity Evaluation: A Fluorescence Microscopy Application

In this chapter, students have a small course on the effects of hydrophobic pollutants on DNA along with the method allowing them to observe and quantify DNA damage. They can find in this chapter all details necessary to perform an experiment on the subject.

16.1   Principle of the Comet Test

The Comet test is a method that allows measurement of the degree of DNA damage within a nucleic cellular population. The general principle of the comet test method can be described as follows:

1   The nucleic cells are lysed in a basic medium inducing a partial unfolding of the chromatin.
2   The cells are stained with a fluorophore.
3   Analysis with a fluorescence microscope of different parameters that characterise the DNA structure.

16.2   DNA Structure

DNA is composed of three chemical functions: A deoxyribose (a pentose, i.e., a sugar with five carbons), organic (nitrogenous) bases (pyrimidines: cytosine and thymine; purines: adenine and guanine), and a phosphoric acid.

Ribose and deoxyribose contribute to the formation of RNA and DNA, respectively. Binding of a base to the pentose yields a nucleoside, and binding of a phosphoric acid to the nucleoside forms a nucleotide. Finally, different nucleotides bound together form a nucleic acid. There is a specific complementary nature to the bases: adenine with thymine and guanine with cytosine. The C–G pair has three hydrogen bonds, while the A–T pair has only two, thus preventing incorrect pairing. The sequence of the bases determines the primary structure of the DNA.

DNA's secondary structure is a helix (Figure 16.1) twisted to the right and known as the DNA B form. The distance between two consecutive bases is 3.4 Å. Since the helix repeats itself approximately every 10 bases, the pitch per turn of the helix is 33.4 Å. B DNA is the major conformation of DNA in solution.

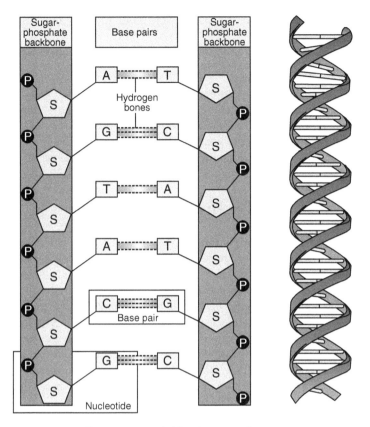

Figure 16.1 DNA structure illustration provided by the National Human Genome Research Institute (http://www.genome.gov/DIR/VIP/Learning_Tools/genetic_illustrations.html).

An alternative conformation of the B DNA appears in solution when the amount of water necessary to hydrate the double helix is not sufficient. In the A conformation, the pitch is 24.6 Å, and a complete turn of the helix needs the presence of 11 base pairs. Finally, we should add that *in vivo*, we do not know whether the A form of DNA does really exist. The third conformation of DNA is the Z form where the helix is twisted toward the left.

The sequence of DNA bases is of great importance in heredity. In fact, "the sequence of bases in one strand has a complementary relationship to the sequence of bases in the other strand." In other terms, information contained in the sequence of one strand is conserved in the second strand.

16.3   DNA Reparation

Damage to DNA, resulting from toxic agents, appears spontaneously and exists permanently. However, most organisms have the ability to repair their DNA. The mechanisms of DNA reparation are generally of three types:

1   Restoration of the damaged zone: An enzymatic reaction restores the initial structure without breaking the skeleton of the molecule.

2   Deletion of the damaged zone: The bases or the group of the damaged nucleotides are deleted then replaced.

3   Tolerance toward the damage zone: In this case, there is no reparation, and the damage is accepted.

Homologous recombination occurs between two homologous DNA molecules. It is also called DNA crossover. It intervenes in the reparation of breaks in DNA resulting mainly from irradiation. In addition, it helps to restore DNA synthesis after blockage of the replication fork. Therefore, homologous recombination plays an important role in maintaining the integrity of the genome. However, in some cases, recombination can generate chromosomal rearrangements harmful to the cell or can induce the formation of toxic intermediates.

16.4   Polycyclic Aromatic Hydrocarbons

Polycyclic aromatic hydrocarbons (PAHs) are persistent pollutants with mutagenic and/or carcinogenic characteristics. They are persistent in the environment because they form complexes with other compounds. They are formed during the incomplete burning of coal, oil, gas, garbage, or other organic substances like tobacco or charbroiled meat. They follow inside the organism different metabolic paths. Some of these paths increase the genotoxic potential of the PAHs inducing the damage of the DNA. PAHs include organic compounds with at least two aromatic rings, and so they are large and flat compounds. Many aromatic compounds, including PAHs, are carcinogenic because of their structure. The flat, hydrophobic shape of a PAH makes it difficult to excrete from the body. In addition, this shape allows a PAH to insert itself into the structure of DNA, where it interferes with the proper functioning of DNA, leading to cancer.

Carcinogenicity can result from inhalation of dangerous products, cutaneous contact, and transfer via the alimentary chain. Eight polycyclic aromatic hydrocarbons, benzo(a) anthracene, chrysene, benzo(a)fluoranthene, benzo(k)fluoranthene, benzo(a)pyrene, dibenzo(ah)anthracene, benzo(ghi)perylene, and indeno(1,2,3-cd)pyrene, are known to be genotoxic via the alimentary chain. In the vertebrate, metabolism of these PAHs will induce the formation of diol epoxydes compounds responsible for carcinogenicity and genotoxicity. For example, benzo(a)pyrenes generate the benzopyrene diol epoxide or BPDE (Figure 16.2).

Benzo(a)pyrene in cigarette smoke coats the lung surface and is absorbed into cells which oxidize benzo(a)pyrene to convert it into a more polar and hence water-soluble form for excretion. The oxidized form intercalates at GC base pairs and modifies guanine bases

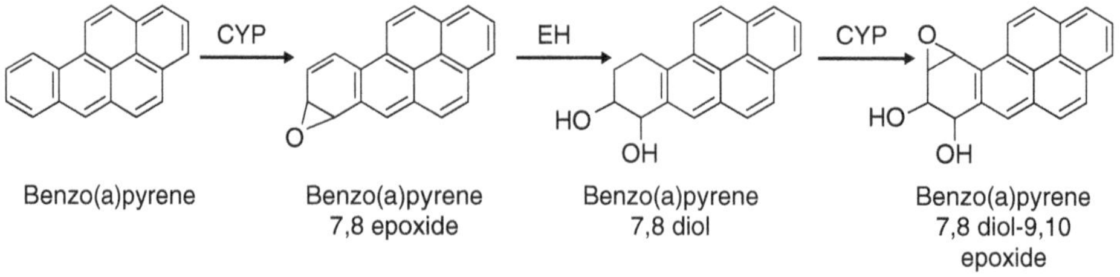

| Benzo(a)pyrene | Benzo(a)pyrene 7,8 epoxide | Benzo(a)pyrene 7,8 diol | Benzo(a)pyrene 7,8 diol-9,10 epoxide |

Figure 16.2   Oxidation of benzo(a)pyrene to BPDE.

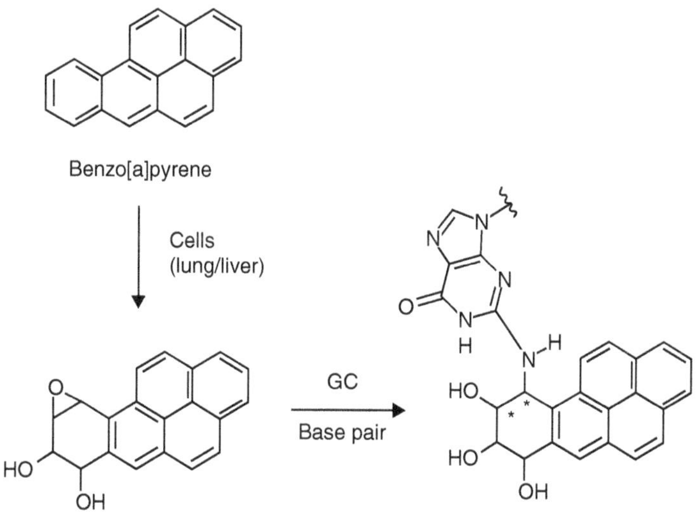

Figure 16.3 Effect of benzo(a)pyrene in cigarette smoke on the guanidine base in the GC sequence.

covalently (Figure 16.3). If the resulting DNA lesion is not repaired, a permanent mutation may be produced when the cells replicate. Mutation can lead to abnormal cell growth and cancer.

16.5 Reactive Oxygen Species

In vivo, in physiological conditions, equilibrium exists between reactive oxygen species production and the antioxidant-defence mechanisms. An antioxidant is a substance that, when present in low concentrations relative to the oxidizable substrate, significantly delays or reduces substrate oxidation (Halliwell 1995). Antioxidants get their name because they fight oxidation. They are substances that protect other chemicals of the body from damaging oxidation reactions by reacting with free radicals and other reactive oxygen species within the body, hence hindering the process of oxidation. During this reaction, an antioxidant sacrifices itself by becoming oxidized. However, the antioxidant supply is not unlimited as one antioxidant molecule can react with only a single free radical. Therefore, there is a constant need to replenish antioxidant resources, whether endogenously or through supplementation (source: Dr. Fouad Tamer for The Doctor's Lounge.net; http://www.thedoctorslounge.net/medlounge/articles/antioxidants/antioxidants1.htm).

Reactive oxygen species are the result of a photochemical reaction of oxygen:

$$O_2 \xrightarrow{\text{Photochemical reaction}}$$

$^1O_2$ singlet oxygen

O_2^- anion superoxide $\rightarrow H_2O_2$ hydrogen peroxide

$OH^\bullet$ hydroxyl radical

Reactive oxygen species can attack all cell compartments, lipids, proteins, and DNA. The cell will activate enzymatic systems to limit the generation of reactive oxygen species. Also, components such as polyphenol, carotenoid, and vitamin C, which are present in food, are good antioxidant molecules.

16.6 Causes of DNA Damage and Biological Consequences

DNA damage can be directly or indirectly generated after exposure to chemical and/or physical environmental agents (UV radiations, ionized radiations, chemical contaminants, etc.). Also, endogenic damage can occur as a result of DNA replication, reparation mechanisms, instability in physiological conditions of some chemical bonds, accumulation or production of reactive metabolites, and generation of reactive oxygen species.

The biological consequences of DNA lesion are: cell death, new mutation as a result of the deficient process of reparation, and induction of a damaged DNA. This will lead to the manifestation of the genotoxic syndrom in the vertebrate and to possible embryo toxicity. Figure 16.4 shows the possible courses followed by a cell after DNA damage.

The effect of DNA toxicity such as cancer, genotoxic disease syndrome, modification of a genetic profile, and abnormalities in development generally appear after several years.

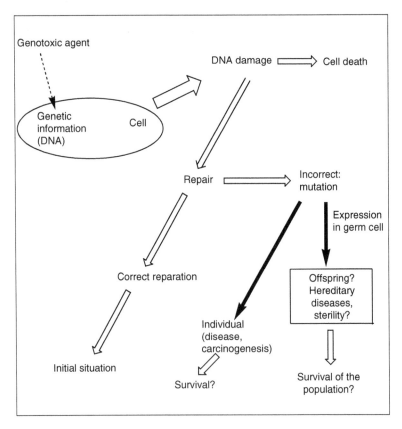

Figure 16.4 Possible courses undergone by a cell after DNA damage. Courtesy of Lemière, S. (2004). Thesis of University of Metz.

16.7   Types of DNA Lesions

16.7.1   Induction of abasic sites, AP, apurinic, or apyrimidinic

AP sites may be spontaneous or induced, and can block the replication. Instability of the glycosylic site constitutes the principal cause of AP sites formation. Cellular DNA can undergo puric base losses, estimated at between 250 and 500 per hour. AP sites can be regenerated during reparation processes by base excision and free radicals.

16.7.2   Base modification

This can occur following one of these different methods:

* Alkylation, by addition of methyl or alkyl groups.
* Deamination, by losing NH_2 groups.
* Hydroxylation and oxidation by the reactive oxygen species and the ionized radiations.
* Dimer formation between pyrimidines, as a result of UV radiation.
* Cross-linking between the two DNA brands or within the same brand or between a DNA brand and a protein, as a result of UV radiation.

16.7.3   DNA adducts

DNA adducts comprise nucleotides where chemical mutagenic substances are covalently bound. The common property of the chemical mutagenic substances is their electrophilic nature. Electrophilic sites (electron deficiency) bind to the nucleophilic sites of DNA or the proteins inducing covalent bonding leading to adduct formation, resulting in DNA conformation distortion, and replication and transcription blockage (Esaka *et al.* 2003).

16.7.4   Simple and double-stranded breaks

Breaks occurring within DNA can be induced by radical-species production. Also, they can result from the intervention of reparation mechanisms of damage to the DNA.

16.8   Principle of Fluorescence Microscopy

Two methods have been developed in fluorescence microscopy: transmitted light fluorescence and epi-illumination fluorescence. In the first method, the specimen is excited by light passing through the condenser lens, and fluorescence emission is captured by the objective lens (Figure 16.5).

As we can see from Figure 16.5, excitation light coming from the condenser lens and going through the objective lens is not completely blocked by the emission filter, thus leading to a high background signal. However, recent developments in filter technology and

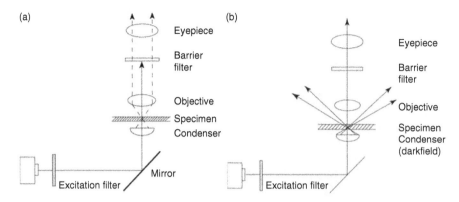

Figure 16.5 (a) Diagram of a fluorescence microscope assembled by the addition of an excitation filter and a barrier filter in a standard bright field microscope. Solid line, excitation light; broken line, emitted light. (b) Diagram of a darkfield fluorescence microscope. The NA of the objective is designed to miss the direct rays from the darkfield condenser (thin lines). Only emitted fluorescence can enter the objective (thick line).

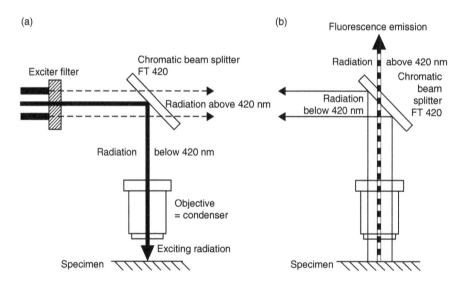

Figure 16.6 Diagram of an incident or epi-illumination fluorescence microscope. (a) Excitation of the spectrum using the objective as a condenser. The chromatic beam splitter or dichroic mirror is designed such that light below a designed cutoff wavelength (here 420 nm) is reflected through the objective (condenser) and excites the specimen (heavy black light). (b) Emission of light from the specimen. Emitted light (heavy dashed line) that enters the objective above the designed cutoff is not reflected, but passes to the detector. Wavelengths below the designed cutoff are reflected away from the detector. Courtesy of Zeiss Instruments (http://www.zeiss.de/de/micro/home_e.nsf).

automated microscope controls have enabled very accurate results to be obtained with the trans-fluorescence method (see, for example, Tran and Chang 2001).

In the epi-illumination method, a dichroic beam splitter comprises two parts: one acting as an excitation filter and the second as an emission filter (Figure 16.6).

16.9   Comet Test

16.9.1   *Experimental protocol*

This test, called the Comet Assay or single-cell gel-electrophoresis assay, allows the degree of DNA damage to be determined within a nucleic cell population. The principle of the method is based on the microelectrophoresis of nuclei of isolated cells, under basic conditions, on agarose gel (the whole being observed under a fluorescence microscope).

1   After exposure, *in vivo* or *in vitro*, eukaryotic cells are surrounded by an agarose gel and lysed in a buffer containing detergents and a high salt concentration.
2   Afterward, DNA is denatured, and a brief electrophoresis is performed under alkaline conditions.
3   After staining with ethidium bromide or any other specific fluorophore, DNA from the intact cell appears as a sphere, while DNA from a damaged cell is stretched progressively to the anode, due to the phosphate ions of DNA, proportional to the number of breaks within the DNA (Figures 16.7 and 16.8). The results are usually displayed as four Comet class types.
4   The lesions can be evaluated in a semi-quatitative manner with the different comet classes or in a quantitative manner with an image analyzer.

In summary, we have the following steps: (1) cell suspension, preparation, viability, and density; (2) cells in agarose gel on slides; (3) lysis; (4) DNA unwinding, electrophoresis, neutralization, and staining with a fluorophore; (5) image analysis and scoring.

Figure 16.9 shows the results from a Comet test of normal, apoptotic, and necrotic cells. One should be careful not to confuse apoptotic and damaged cells.

16.9.2   *Nature of damage revealed with the Comet test*

- Up to pH 9, double-stranded breaks are observed.
- Up to pH 12.1, certain double- and single-stranded breaks are observed.
- Up to pH 13.0, double- and single-stranded breaks are observed along with alkali label sites.

16.9.3   *Advantages and limits of the method*

- Approach cell by cell.
- Studies can be performed on all cell types.
- The method is fast and flexible.
- A few cells can be studied at the same time.
- The method is very sensitive. In fact, it allows the detection of 100–1000 breaks per nucleus. One can estimate 0.1–1 break per 10^9 Da, and one can detect DNA damage as a result of 50 mGy X-rays.
- This method allows the detection of single- and double-stranded breaks and alkali labile sites. These breaks can be a result of radiation, direct and indirect mutagens, alkylation agents, intercalating agents, oxidative DNA damage, and incomplete DNA repair.

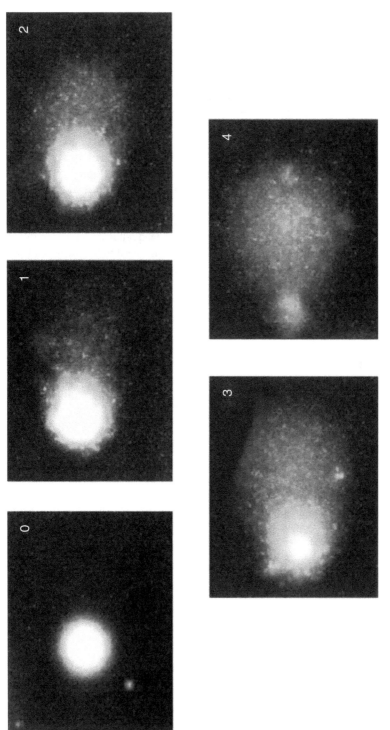

Figure 16.7 Images of comets (from lymphocytes), stained with DAPI. These represent classes 0–4, as used for visual scoring. Source: Collins, A.R. (2004). *Molecular Biotechnology*, **26**, 249–261. Reprinted with permission from Humana Press Inc. Reproduced in Color plate 16.7.

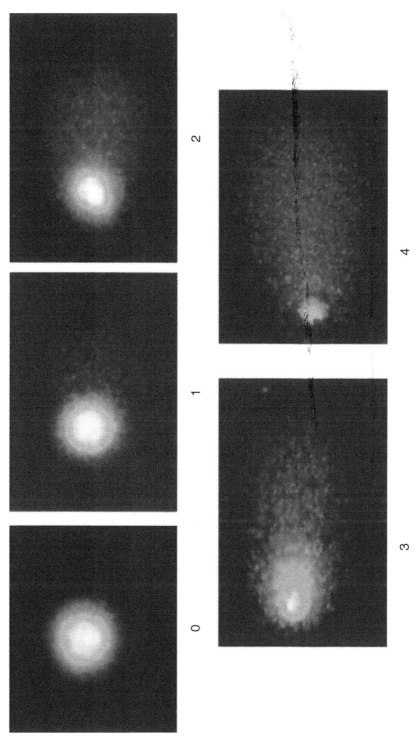

Figure 16.8 CHO cells treated with ethyl methanesulfonate; Magnification 350×. Courtesy of Dr. Alok Dhawan, Industrial Toxicology Research Centre, Mahatma Gandhi Marg, Lucknow, UP. www.cometassayindia.org and www.cometassayindia.org/protocols.htm. Reproduced in Color plate 16.8.

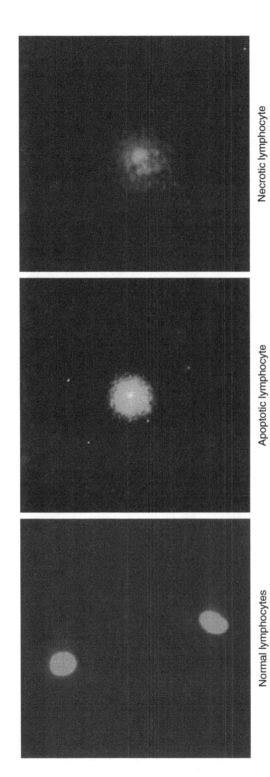

Normal lymphocytes   Apoptotic lymphocyte   Necrotic lymphocyte

Figure 16.9 Comet test for lymphocyte cells in three states: normal, apoptotic, and necrotic. The reader can refer to the following sites: www.cometassayindia.org and www.cometassayindia.org/protocols.htm. Reproduced in Color plate 16.9.

16.9.4   Result expression

DNA damage observed via the Comet test is expressed as follows:

- Percentage of DNA in the comet tail.
- Length of the tail.
- Extent of tail moment, which characterizes the product of % DNA in the tail with the length of the tail.
- Olive tail moment, which is equal to the product of the % DNA in the tail with the distance between the gravity center of the tail and that of the head of the comet, calculated from the fluorescence intensity.

$$\% \text{ DNA in the comet head} = \frac{I_F \text{ in the head}}{I_F \text{ in the head} + I_F \text{ in the tail}} \times 100$$

where I_F is the fluorescence intensity.

References

Collins, A.R. (2004). The comet assay for DNA damage and repair. Principles, applications and limitations. *Molecular Biotechnology*, **26**, 249–261.

Esaka, Y., Inagaki, S. and Goto, M. (2003). Separation procedures capable of revealing DNA adducts. *Journal of Chromatography B – Analytical Technologies in the Biomedical and Life Sciences*, **797**, 321–329.

Halliwell, B. (1995). Antioxydants. *The Biochemist*, **17**, 3–6.

Lemière, S. (2004). Interest of the Comet Assay for the study of environmental genotoxicity. Thesis, University of Metz.

Tran, P.T. and Chang, F. (2001). Transmitted light fluorescence microscopy revisited. *The Biological Bulletin*, **201**, 235–236.

www.cometassayindia.org

www.cometassayindia.org/protocols.htm

Chapter 17
Questions and Exercises

17.1 Questions

17.1.1 Questions with short answers

1 Describe the correlation that may exist between the Morse curve and the energy potential curve of an anharmonic oscillator.
2 What are the parameters that influence the intensity of an absorption band?
3 Explain why it is important to record fluorescence spectra at low product concentrations.
4 Describe the inner filter effect.
5 What are the advantage(s) and the disadvantage(s) of using a laser instead of a lamp as an excitation wavelength?
6 Describe effects of temperature on the fluorescence intensity, anisotropy, and lifetime.
7 Explain how we can use fluorescence tyrosine to follow conformational changes within a protein.
8 What is the common property shared by fluorescence and phosphorescence emission?
9 What are the difference(s) that may exist between $T_1 \rightarrow S_0$ and $T_1 \rightarrow S_1$ transitions?
10 What are the difference(s) that may exist between emission from a nonrelaxed state and emission from a non-polar medium?

17.1.2 Find the error

Five of these sentences are wrong. Write them after doing the appropriate corrections:

a The passage of an electron from the excited state T_1 to the excited state S_1 is called intersystem crossing.
b The phosphorescence lifetime goes from the ns to the ms range.
c The fluorescence quantum yield of a fluorophore is independent of the excitation wavelength.
d Fluorescence emission at short wavelengths characterizes a fluorophore in a nonrelaxed and apolar medium.

e   Fluorophore molecules dissolved in an apolar medium are associated to the solvent molecules by their dipoles.

f   The multiple fluorescence lifetimes of a fluorophore are dependent on different factors such as the nonrelaxed states of the fluorophore.

17.1.3   Explain

a   We have recorded the fluorescence emission spectrum of Trp residues of a protein at a λ_{ex} of 295 nm using different slits at the excitation and emission wavelengths, and obtained the spectra shown in Figure 17.1. Regardless of the intensity variation, do you observe any significant difference between the recorded spectra?

b   Why does an increase in temperature broaden the spectral bands?

c   How can we know whether a solution contains one or two fluorophores?

d   The principle of resonance energy transfer.

e   The error we can induce if we record the fluorescence spectra of the blank and the fluorophore solution at two different temperatures.

f   The principle of studying molecules diffusion in cells with the method of photobleaching.

g   Why does an increase in temperature decrease the fluorescence quantum yield?

h   Why might the information obtained from dynamic fluorescence quenching be different from that obtained with dynamic phosphorescence quenching?

i   How can you explain the existence of a fluorescence quantum yield?

j   What are the difference(s) between isobestic and isoemissive points?

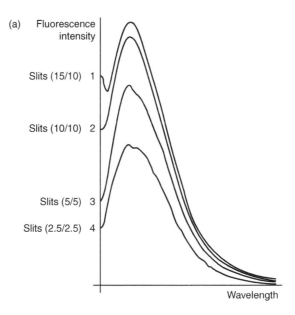

Figure 17.1   Fluorescence emission spectrum of Trp residues within a protein recoded at different excitation/emission slits.

17.1.4 Exercises

17.1.4.1 Exercise 1

We have determined the rotational correlation time and the fluorescence intensity of the unique Trp residue of a protein, at the native and the denatured states at three temperatures, 5, 17, and 44°C and obtained the following values:

Φ_c (ns)	0.1	5	10	2	1	15
I (au)	5	7	10	20	17	15

Assign to each value of Φ_c and I the temperature of the measurement and the state of the protein. Explain your assignment.

17.1.4.2 Exercise 2

The heme optical density of a hemoprotein and the fluorescence intensity of the unique Trp residue of the protein are recorded at hydrostatic pressures between 0 and 4 kbar. The two recorded parameters increase with hydrostatic pressure. Explain in five to seven lines.

17.1.4.3 Exercise 3

We have determined the emission maximum of TNS bound to a protein at two hydrostatic pressures (0 and 5 kbar) and at three different temperatures, −60, 0, and 40°C ($\lambda_{ex} = 360$ nm), and obtained the following results:

	0 kbar	5 kbar
−60°C	420 nm	425 nm
0°C	430 nm	435 nm
40°C	433 nm	435 nm

Can you explain the results obtained?

17.1.4.4 Exercise 4

We have determined the fluorescence polarization P and the bimolecular diffusion constant k_q of oxygen of the unique Trp residue of a protein at three temperatures, 5, 20, and 30°C, and at two hydrostatic pressures (0–4 kbar), and obtained the following results:

P	0.015	0.03	0.05	0.1	0.2	0.3
k_q (M^{-1} s^{-1})	0.5×10^8	4×10^8	8×10^8	1×10^9	5×10^9	9×10^9

Assign to each value of P and k_q the corresponding temperature and hydrostatic pressure. Explain your assignment in five to seven lines.

17.1.4.5 Exercise 5

The polarization and the rotational correlation time of the unique Trp residue of a protein have been measured at two hydrostatic pressures, 0 and 8 kbar, at three temperatures, 5, 17, and 44°C. The following values are obtained:

P	0.1	0.06	0.05	0.3	0.2	0.44
Φ_c	2.9	8.5	44.6	1.5	4.4	22.4

Assign to each value of P and Φ_c the corresponding temperature and hydrostatic pressure. Explain your assignment in a maximum of seven lines.

17.1.4.6 Exercise 6

The fluorescence lifetime and polarization of an extrinsic probe bound tightly to a spherical protein have been measured at different oxygen concentrations. The following data are obtained:

O_2 (mM)	P	τ (ns)
0.218	0.156	15
0.250	0.178	12
0.450	0.208	9
0.700	0.250	6

What is the molecular weight of the protein? The viscosity η at 20°C is 0.01 g cm^{-1} s^{-1}. The Boltzman constant $k = 1.3805 \times 10^{-6}$ g cm^2 s^{-2} K^{-1}. The Avogadro number $N = 6.022 \times 10^{23}$.

17.1.4.7 Exercise 7

The fluorescence intensity and anisotropy of the flavin of a flavoprotein have been measured from 0 to 4 kbar. We notice an increase in the fluorescence intensity and a decrease in the fluorescence anisotropy. Explain the results in five to seven lines.

17.1.4.8 Exercise 8

We have determined the fluorescence intensity and lifetime of the unique Trp residue of a protein, at the native and denatured states at three temperatures, 5, 17, and 44°C, and obtained the following values:

τ	0.1	0.3	0.5	1	3	5 ns
I	5	10	20	40	80	100 au

Assign to each value of τ and I the temperature of the measurement and the state of the protein. Explain your assignment.

17.1.4.9 Exercise 9

Describe a fluorescent method allowing the conformational change of a protein to be followed with the two probes: Trp residue and TNS.

17.1.4.10 Exercise 10

1 The fluorescence polarization of the unique Trp residue of a monomeric protein has been measured as a function of the temperature over viscosity ratio. The following values have been obtained:

P	0.200	0.190	0.182	0.174
T/η	190	260	340	410

Knowing that the value of the limiting polarization at the excitation wavelength is 0.24, the rotational correlation time of the protein is 20 ns, and the fluorescence lifetime of the Trp residue is 2.8 ns, can you tell whether the fluorophore presents residual motions or not?

2 If we measure the anisotropy decay of the Trp residue of the protein as a function of time, based on the answer you found in question 1, what would be the expected result?

3 The fluorescence intensity of the Trp residue of the protein is measured as a function of temperature, what will be the result that we should obtain and why? The viscosity η at 20°C is 0.01 g cm^{-1} s^{-1}. The Boltzman constant $k = 1.3805 \times 10^{-6}$ g cm^2 s^{-2} K^{-1}. The Avogadro number $N = 6.022 \times 10^{23}$.

17.1.4.11 Exercise 11

Binding experiments have been performed between the fluorophore TNS and a flavodeshydrogenase extracted from the yeast *Hansenula anomala*. The fluorescence spectrum of the Trp residues of the protein was recorded from 320 to 370 nm, in the absence and presence of different concentrations of TNS. Two sets of experiments were carried out, one in phosphate buffer pH 7 and one in 6 M guanidine, pH 7. The following fluorescence intensities (I_F) were obtained:

Experiment 1

[TNS] $= 0\ \mu$M		[TNS] $= 0.74\ \mu$M		[TNS] $= 0.95\ \mu$M	
λ (nm)	I_F	λ (nm)	I_F	λ (nm)	I_F
320	14.2	320	13.3	320	12.5
325	16.5	325	15.5	325	14.7
330	18.3	330	17.4	330	16.5
335	19.3	335	18.5	335	17.6
340	19.6	340	18.7	340	17.9
342	19.4	342	18.6	342	17.8
343.5	19.2	343.5	18.5	343.5	17.7
345	19.0	345	18.3	345	17.5
350	17.7	350	17.1	350	16.4
355	16.1	355	15.5	355	14.9
360	14.3	360	13.8	360	13.2
365	12.4	365	12	365	11.5
370	10.8	370	10.5	370	10.2

[TNS] $= 2.33\ \mu$M		[TNS] $= 3.17\ \mu$M		[TNS] $= 3.97\ \mu$M	
λ (nm)	I_F	λ (nm)	I_F	λ (nm)	I_F
320	11.8	320	10.8	320	10.5
325	13.8	325	12.7	325	12.2
330	15.6	330	14.5	330	14
335	16.8	335	15.6	335	15.2
340	17.3	340	16	340	15.8
342	17.2	342	15.9	342	15.8
343.5	17.1	343.5	15.9	343.5	15.7
345	16.9	345	15.7	345	15.6
350	15.9	350	14.75	350	14.75
355	14.5	355	13.5	355	13.4
360	13.0	360	12.1	360	12.1
365	11.3	365	10.5	365	10.4
370	10	370	9.4	370	9.3

[TNS] = 4.45 μM		[TNS] = 5.18 μM		[TNS] = 6.14 μM	
λ (nm)	I_F	λ (nm)	I_F	λ (nm)	I_F
320	10	320	9.7	320	9.2
325	11.8	325	11.5	325	10.8
330	13.6	330	13.2	330	12.6
335	14.9	335	14.4	335	13.8
340	15.3	340	14.9	340	14.3
342	15.4	342	15.1	342	14.4
343.5	15.4	343.5	15.0	343.5	14.4
345	15.2	345	14.8	345	14.3
350	14.5	350	14.1	350	13.65
355	13.3	355	12.9	355	12.5
360	12.1	360	11.5	360	11.2
365	10.3	365	10.1	365	9.8
370	9.2	370	8.9	370	8.7

[TNS] = 6.56 μM		[TNS] = 6.98 μM		[TNS] = 7.88 μM	
λ (nm)	I_F	λ (nm)	I_F	λ (nm)	I_F
320	8.7	320	8.3	320	8.0
325	10.3	325	9.7	325	9.4
330	12.0	330	11.4	330	11.0
335	13.2	335	12.6	335	12.3
340	13.8	340	13.3	340	13.0
342	13.8	342	13.35	342	13.0
343.5	14.0	343.5	13.45	343.5	13.1
345	13.8	345	13.4	345	13.1
350	13.2	350	12.7	350	12.5
355	12.2	355	11.8	355	11.5
360	10.8	360	10.6	360	10.4
365	9.5	365	9.3	365	9.2
370	8.5	370	8.3	370	8.2

[TNS] = 8.68 μM		[TNS] = 9.42 μM	
λ (nm)	I_F	λ (nm)	I_F
320	7.6	320	7.1
325	9.0	325	8.5
330	10.6	330	10.1
335	11.3	335	11.3
340	12.6	340	12
342	12.7	342	12.1
343.5	12.8	343.5	12.3
345	12.7	345	12.2
350	12.2	350	11.8
355	11.3	355	11.0
360	10.2	360	10.0
365	8.9	365	8.9
370	8.1	370	8.0

Experiment 2

[TNS] = 0 μM		[TNS] = 2 μM		[TNS] = 4 μM	
λ (nm)	I_F	λ (nm)	I_F	λ (nm)	I_F
320	7.2	320	7.0	320	6.7
325	8.8	325	8.2	325	7.8
330	10.7	330	10.0	330	9.7
335	12.8	335	12.0	335	11.7
340	14.3	340	13.6	340	13.3
345	16.2	345	15.6	345	15.2
350	17.2	350	16.7	350	16.3
355	17.4	355	17.1	355	16.6
360	17.0	360	16.9	360	16.4
365	16.0	365	16.0	365	15.5
370	15.0	370	15.0	370	14.6

[TNS] = 8 μM		[TNS] = 12 μM	
λ (nm)	I_F	λ (nm)	I_F
320	6.6	320	6.6
325	7.7	325	7.7
330	9.5	330	9.5
335	11.5	335	11.5
340	13.0	340	13.0
345	15.0	345	15.0
350	16.1	350	16.1
355	16.4	355	16.4
360	16.1	360	16.1
365	15.3	365	15.3
370	14.4	370	14.4

1 Plot the spectra of each experiment.
2 Can you tell from the data and from the spectra obtained, the buffer in which each experiment was carried on? Explain your answer.
3 Can you give a logical explanation of the slight shift observed in the first experiment?
4 Why is this shift not observed in the second experiment?
5 The data shown before are from spectra not corrected for the inner filter effect. The optical density measured at the excitation wavelength (280 nm) and at two emission wavelengths, 335 for experiment 1 and 355 for experiment 2, gives the following values:

Experiment 1

[TNS] (μM)	Volume in the cuvette (μl)	OD λ_{ex} = 280 nm	λ_{em} = 335 nm
0	1060	0.086	0.008
0.74	1061.5	0.092	0.016
0.95	1063	0.107	0.026
2.33	1064.5	0.116	0.030
3.17	1066	0.130	0.040
3.97	1067.5	0.143	0.047
4.45	1069	0.151	0.052
5.18	1070.5	0.162	0.059
6.14	1072	0.181	0.067
6.56	1073.5	0.190	0.072
6.98	1075	0.200	0.078
7.88	1076.5	0.210	0.084
8.68	1078	0.223	0.092
9.42	1079.5	0.234	0.099

Experiment 2

[TNS] (μM)	Volume in the cuvette (μl)	OD $\lambda_{ex} = 280$ nm	$\lambda_{em} = 335$ nm
0	1100	0.090	0.012
2	1110	0.120	0.034
4	1120	0.147	0.045
6	1130	0.185	0.058
8	1140	0.214	0.070

Correct the fluorescence intensities at 335 and 355 nm for the optical densities. What can you conclude?

The optical densities measured with a spectrophotometer were obtained with an optical path length equal to 1 cm. However, optical path lengths equal to 1 and 0.4 cm were used for the emission and excitation wavelengths, respectively.

17.1.4.12   Exercise 12

O-Nitrophenyl glactoside hydrolysis with β-galactosidase in the absence and presence of three inhibitors, *O*-nitrophenyl β-D-thiogalactoside (ONPTG), maltose, and melibiose, was studied at 25°C in pH 7.5 buffer. Table 17.1 shows the optical densities at 410 nm of the PNP formed, expressed per minute.

Table 17.1   Optical densities at 410 nm of PNP obtained by minute.

[Substrate] (M)	v_0	v_1 (3×10^{-4} M of ONPTG)	v_2 (0.26 M of maltose)	v_3 (0.17 M of melibiose)
2.5×10^{-5}	0.033	0.018	0.0165	0.027
5×10^{-5}	0.055	0.033	0.0275	0.041
1×10^{-4}	0.0825	0.055	0.041	0.055
2.5×10^{-4}	0.118	0.091	0.059	0.069
5×10^{-4}	0.138	0.118	0.069	0.075
1×10^{-3}	0.15	0.138	0.075	0.079

Plot $v = f([S])$ and determine the types of inhibition. Calculate the kinetic parameters.

17.2   Solutions

17.2.1   Questions with short answers

1   *Describe the correlation that may exist between the Morse curve and the energy potential curve of a anharmonic oscillator.*

R1. Within a molecule, the stretching of the molecular bond lets the atoms come closer and move away one from each other. The Morse curve describes the potential energy induced by this stretching. At short distances, the repulsive forces are dominant, while at long distances, the attractive forces are dominant. However, when the distance that separates the two atoms within the bond is too important, the attractive forces are no longer efficient, and the Morse curve will look like an anharmonic oscillator.

2   *What are the parameters that influence the intensity of an absorption band?*
   R2. The position of the nuclei at the moment of the electronic transition and the population of molecules that have reached the excited state.

3   *Explain why it is important to record fluorescence spectra at low product concentrations.*
   R3. A high concentration may increase the optical density at the excitation and/or the emission wavelengths. This will distort the fluorescence emission spectrum by decreasing the real fluorescence intensity and by shifting the emission peak.

4   *Describe the inner filter effect.*
   R4. At a high optical density at the excitation and/or emission wavelengths, a distortion of the fluorescence emission spectrum is observed. A fluorescence intensity decrease is observed, and the emission peak is shifted.

5   *What are the advantage(s) and the disadvantage(s) of using a laser instead of a lamp as an excitation source?*
   R5. The energy of a laser is higher than that of a lamp, thus allowing one to work at low fluorophore concentrations. However, since fluorophores are light-sensitive, excitation with a laser could accelerate the photo-bleaching rate.

6   *Describe the temperature effects on fluorescence intensity, anisotropy, and lifetime.*
   R6. Temperature favors the deactivation processes other than fluorescence and so decreases the fluorescence intensity, quantum yield, and lifetime. Temperature increases the local motions of the fluorophores and so decreases the value of the polarization.

7   *Explain how can we use tyrosine fluorescence to follow conformational changes within a protein.*
   R7. Fluorescence of tyrosine in protein is generally quenched as a result of the tertiary structure of the protein. A structural modification of the protein would affect the fluorescence properties of the protein, including those of the tyrosine. Thus, in some cases, one can observe an increase in the fluorescence intensity at 303 nm and the quantum yield of the tyrosine and a decrease in its anisotropy.

8   *What is the common property shared by fluorescence and phosphorescence emission?*
   R8. Photon emission.

9   *What are the difference(s) that may exist between $T_1 \rightarrow S_0$ and $T_1 \rightarrow S_1$ transitions?*
   R9. $T_1 s \rightarrow S_0$ transition is possible since T_1 is energetically higher than S_0.
   $T_1 \rightarrow S_1$ transition is not possible because T_1 is energetically lower than S_1.

10 *What are the difference(s) that may exist between emission from a nonrelaxed state and emission from a non-polar medium?*
   R10. Emission from a nonrelaxed state means that it occurs before the dipole reorientation of the solvent molecules or the dipole of the amino acids of the fluorophore binding site. Emission from a non-polar medium means that the fluorescence occurred after the inducing by the fluorophore of a dipole in the medium.

17.2.2 Find the error

Five of these sentences are wrong. Write them out after doing the appropriate corrections:

1   *The passage of an electron from the excited state T_1 to the excited state S_1 is called intersystem crossing.*
    R1. The passage of an electron from the excited state S_1 to the excited state T_1 is called intersystem crossing.
2   *The phosphorescence lifetime goes from the ns to the ms range.*
    R2. The phosphorescence lifetime stands from the ms to the s, min, or h range.
3   *The fluorescence quantum yield of a fluorophore is independent of the excitation wavelength.*
    R3. The sentence is correct.
4   *Fluorescence emission at short wavelengths characterizes a fluorophore in a nonrelaxed and apolar medium.*
    R4. Fluorescence emission at short wavelengths characterizes a fluorophore in a nonrelaxed or non-polar medium.
5   *Fluorophore molecules dissolved in a non-polar medium are associated to the solvent molecules by their dipoles.*
    R5. Fluorophore molecules dissolved in a polar medium are associated to the solvent molecules by their dipoles.
    or
    The dipoles of fluorophore molecules dissolved in a polar medium are associated with the dipoles of the solvent molecules.
    or
    The dipoles of fluorophore molecules dissolved in a non-polar medium are associated with the induced dipoles of the solvent molecules.
6   *The multiple fluorescence lifetimes of a fluorophore are dependent on different factors such as the nonrelaxed states of the fluorophore.*
    R6. The multiple fluorescence lifetimes of a fluorophore are dependent on different factors such as the nonrelaxed states of the fluorophore emission.

17.2.3 Explain

a   *At large slits, vibrational transitions are not observed on the emission spectra.*
b   *Why does a temperature increase broaden the spectral bands?*
    Rb. Temperature increases the frequency of molecular collisions. The vibrational transitions of the nuclei are badly resolved. This will induce a broadening of the spectral bands.
c   *How can we know whether a solution contains one or two fluorophores?*
    Rc. The fluorescence emission of a solution containing two fluorophores varies with excitation wavelength, which is not the case in the presence of one fluorophore.
d   *The principle of resonance energy transfer.*
    Rd. The electron of the excited molecules induces an oscillating electric field that excites the electrons of the acceptor molecules.

e *The error we can induce if we record the fluorescence spectra of the blank and the fluorophore solution at two different temperatures.*

   Re. In principle, the blank is not fluorescing. However, one should subtract the Raman spectrum of the blank from the recorded fluorescence spectrum of the fluorophore. Therefore, the temperature should be the same for both spectra; otherwise, an error could be induced in determining the real intensity of the fluorophore emission.

f *The principle of studying molecules diffusion in cells with the method of photo bleaching.*

   Rf. We induce an irreversible photo-bleaching of fluorophore molecules within a specific region of the cell. The appearance of fluorescence in this area after a short time is the result of fluorophore diffusion from other cell areas to the irradiated area.

g *Why does an increase in temperature decrease the fluorescence quantum yield?*

   Rg. Increasing the temperature will induce an increase in the Brownian motions and the intramolecular fluctuations. Therefore, deactivation of the excited singlet state S_1 via nonradiative processes will be favored, and there will be fewer emitted photons. This induces a decrease in the quantum yield.

h *Why might the information obtained from dynamic fluorescence quenching be different from that obtained with dynamic phosphorescence quenching?*

   Rh. Since the phosphorescence lifetime is longer than the fluorescence lifetime, the information obtained is within two different timescales and so can be different.

i *How can you explain the existence of a fluorescence quantum yield?*

   Ri. Since different processes deactivate the excited state of the molecule, there are fewer photons emitted via fluorescence than photons originally absorbed by the fluorophore.

j *What are the difference(s) that exist between isobestic and isoemissive points?*

   Rj. We use the term isobestic when we are dealing with absorption spectroscopy and the term isoemissive when we are dealing with fluorescence spectroscopy.

17.2.4 Exercises solutions

17.2.4.1 Exercise 1

At a specific temperature, in the native state, the protein is more compact than in the denatured state. Thus, the motion of the fluorophore in the native state is slower than its motion in the denatured state. Therefore, the rotational correlation time of the fluorophore is longer in the native state.

For a specific protein state, increasing the temperature will accelerate the motion of the fluorophore. Therefore, the rotational correlation time will be shorter when the temperature is higher.

In the native state, the compactness of the protein increases the interaction between the fluorophore and the surrounding amino acids, thereby decreasing the fluorescence intensity. Thus, the fluorescence intensity should be lower in the native state than when the protein is denatured.

Increasing the temperature will facilitate deactivation of the fluorophore via the nonradiative process and so will decrease the temperature.

	Native state		Denatured state	
5°C	10 au	2 ns	20 au	15 ns
17°C	7 au	1 ns	17 au	10 ns
44°C	5 au	0.1 ns	15 au	5 ns

17.2.4.2 Exercise 2

At 0 kbar, the protein is not denatured but disrupted. Hydrostatic pressure favors heme displacement from its binding site, the heme pocket. Hemin (free heme) will absorb more than bound heme. In the absence of heme in the protein, energy transfer from Trp residues to heme decreases, thereby inducing an increase in the fluorescence intensity.

17.2.4.3 Exercise 3

TNS fluorescence will characterize the state of the protein. At 5 kbar, the protein is not yet denatured, although its tertiary structure is stabilized.

At low temperatures, TNS emission will occur from a nonrelaxed state, and the emission peak will be blue-shifted compared to the spectrum obtained at high temperatures. This is why there is a shift of 10 nm between the spectra recorded at −60 and 0°C.

In the native state, the shift observed between 0 and 40°C is 3 nm, indicating that the structure around TNS is not loose. Hydrostatic pressure destabilizes the tertiary structure of the protein. In this case, between 0 and 40°C, there is no shift in the emission peak.

17.2.4.4 Exercise 4

At 0 kbar, the protein is more compact than at 4 kbar. Thus, motions and fluctuations within the protein are weaker at 0 kbar. Therefore, for the same temperature, P is higher at 0 kbar than at 4 kbar, and k_q is weaker.

When the temperature increases, protein motions increase, accompanied by a decrease in the value of P. Also, oxygen diffusion increases, accompanied by an increase in k_q value.

	0 kbar		4 kbar	
	P	k_q	P	k_q
5°C	0.3	0.5×10^8	0.05	1×10^9
20°C	0.2	4×10^8	0.03	5×10^9
30°C	0.1	8×10^8	0.01	9×10^9

17.2.4.5 Exercise 5

At low temperatures, motions are weak, and the rotational time is high. Also, weak motions induce a high polarization value P.

At 0 kbar, the tertiary structure of the protein is conserved. However, at 8 kbar, the protein is denatured, and so for the same temperature, the protein will have a higher degree of motion at 8 kbar than at 0 kbar:

	0 kbar			8 kbar		
t (°C)	5	17	44	5	17	44
P	0.44	0.3	0.2	0.1	0.06	0.05
Φ_c (ns)	44.6	22.4	8.5	4.4	2.9	1.5

17.2.4.6 Exercise 6

The fluorophore is bound tightly to the protein, so it will follow the motions of the protein. Therefore, the rotational correlation time calculated from the Perrin plot will be equal to the global rotational time of the protein.

Plotting $1/P$ vs. τ yields a linear plot with a slope equal to $(1/P_o - 1/3) \times 1/\Phi_P$ (Figure 17.2), where Φ_P is the rotational correlation time of the protein. Upon extrapolation to $T/\eta = 0$, we obtain the value of P_o equal to $0.416(1/P_o = 2.4)$.

The slope of the plot is 0.2667. This yields a value for Φ_P equal to 7.76 ns. The molecular mass of the protein can be calculated from the following equation:

$$\Phi_P = M(v + h)\eta/kTN \tag{17.1}$$

where M is the molecular mass, $v = 0.73$ cm³ g⁻¹ characterizes the specific volume, $h = 0.3$ cm³ g⁻¹ is the hydration degree, η is the medium viscosity, and N is the Avogadro number.

Thus

$$7.76 = \frac{M \times 1.01 \times 0.01}{1.3805 \times 10^{-16} \times 293 \times 6.022 \times 10^{23}}$$

$M = 18715.$

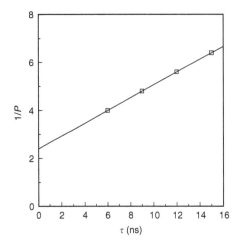

Figure 17.2 Perrin plot of the DNS–protein complex.

17.2.4.7 *Exercise 7*

Up to 4 kbar, protein is not denatured. However, slight structural changes can be observed, inducing flavin release from its binding site. Free flavin in solution has almost no interactions with the protein, and so energy transfer between the flavin and the amino acids of its binding site decreases, which induces an increase in the fluorescence intensity of the co-factor.

The motions of free flavin in solution are much more important than those of bound flavin. Thus, the anisotropy of flavin will decrease with increasing hydrostatic pressure.

17.2.4.8 *Exercise 8*

The tertiary structure of the protein induces a high energy transfer between the Trp residue and the neighboring amino acids. This induces low values in the fluorescence parameters of the Trp residue in the native state in comparison to those measured when the protein is denatured, and so the energy transfer is weak.

A temperature increase induces energy loss via the nonradiative process, thereby decreasing the values of the fluorescence parameters.

	Native state		Denatured state	
	I_F	τ	I_F	τ
5°C	20	0.5	100	5
17°C	10	0.3	80	3
44°C	5	0.1	40	1

17.2.4.9 *Exercise 9*

A Trp residue embedded in the protein matrix will in principle fluoresce with a maximum located at 330 nm. A conformational change may induce a displacement of the Trp residue toward the protein surface. Thus, a shift of the maximum peak to the highest wavelengths (340–350 nm) will be observed.

A structural modification of the protein could induce the displacement of the TNS from its binding site and so the fluorescence intensity decrease in the fluorophore.

17.2.4.10 *Exercise 10*

Let us plot the Perrin plot $1/P$ vs. T/η and calculate the rotational correlation time Φ_c and the polarization $P(0)$ at $T/\eta = 0$. If Φ_c is equal to that of the protein and $P(0)$ to the limiting polarization, then the fluorophore is bound tightly to the protein. Otherwise, the fluorophore displays free and independent motions.

17.2.4.11   Exercise 11

1.

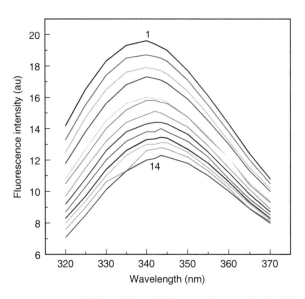

Figure 17.3   Fluorescence spectra of Trp residues of *Hansenula anomala* flavodehydrogenase, in the absence (spectrum 1) and presence (spectra 2–14) with increasing concentrations of TNS. The spectra obtained are from Experiment 1.

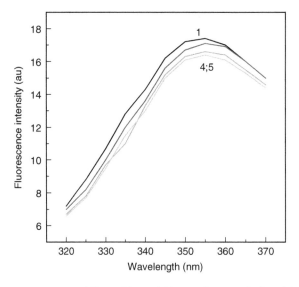

Figure 17.4   Fluorescence spectra of Trp residues of *Hansenula anomala* flavodehydrogenase, in the absence (spectrum 1) and presence (spectra 2–5) of increasing concentrations of TNS. The spectra obtained are from Experiment 2.

2   Experiment 1 was carried out in phosphate buffer, pH 7. In fact, the fluorescence maximum is located at 340 nm, indicating a fluorescence arising from a structured protein. Experiment 2 was carried out in Guanidine, since the Trp residues emit from a denatured state (355 nm). Guanidine denatures proteins.

3   The slight shift to the red observed in the presence of TNS could be simply the result of the inner filter effect due to the absorption of TNS, or it may indicate a slight conformational change of the protein around the Trp residues. In order to find out which of the two cases we have here, we have to replot the spectra after correcting the intensities for the inner-filter effect. If the peaks of the corrected spectra shift to 340 nm, the red-shift observed is the result of the inner filter effect. However, if the positions of the peak do not change, the red shift observed is the result of a slight conformational change around the Trp residues.

4   In the presence of 6 M guanidine, the protein is completely denatured. Trp residues are fully exposed to the solvent. In this case, the interaction between the protein and TNS is weak, almost non-existent. Therefore, the addition of TNS to the protein will not affect the fluorescence spectrum of the Trp residues of the protein.

5   First, we should remember that the fluorescence intensities shown are corrected from the background intensities of the buffer solution or from guanidine solution. Therefore, observed fluorescence intensities should be corrected for the dilution, and finally corrections are made for the absorption at the excitation and emission wavelengths using the following equation:

$$F_{corr} = F_{vol} \times 10^{[OD(em)+OD(ex)]/2}$$

where F_{corr} is the corrected fluorescence intensity for the inner filter effect, and $OD_{(em)}$ and $OD_{(ex)}$ are the optical densities at the excitation and emission wavelengths, respectively.

Fluorescence measurement of the Trp residues of the protein in the absence of TNS was performed in a total volume equal to 1060 and 1100 for the native and denatured proteins, respectively. Addition of TNS increases the initial volume and so dilutes the protein solution. Therefore, in order to see the real effect of TNS addition on the protein fluorescence, we should correct the fluorescence intensities for the dilution effect. In the present experiments, we can see that the volume of TNS added to the protein solution is small, and so corrections should have only a minor effect. However, let us do these corrections using the following equation:

$$F_{vol} = \frac{F_{rec} \times \text{volume in the presence of TNS}}{\text{volume in the absence of TNS}}$$

Experiment 1

Correction for the volume

[TNS] (μM)	Volume in the cuvette (μl)	I_{rec}	I_{vol}
0	1060	19.3	19.3
0.74	1061.5	18.5	18.53
0.95	1063	17.6	17.65
2.33	1064.5	16.8	16.87
3.17	1066	15.6	15.688
3.97	1067.5	15.2	15.30
4.45	1069	14.9	15.03
5.18	1070.5	14.4	14.54
6.14	1072	13.8	13.96
6.56	1073.5	13.2	13.37
6.98	1075	12.6	12.78
7.88	1076.5	12.3	12.49
8.68	1078	11.3	11.49
9.42	1079.5	11.3	11.51

Correction for the inner filter effect

The optical densities are obtained with a spectrophotometer using a path length equal to 1 cm. However, when fluorescence experiments were performed, the optical path length upon excitation was equal to 0.4 cm. Thus, the real optical density at the excitation wavelength is equal to that measured on the spectrophotometer divided by 2.5.

[TNS] (μM)	$\lambda_{ex} = 280$ nm $OD_{(experiment)}$	$\lambda_{em} = 335$ nm $OD_{(experiment)}$	f_c	I_{vol}	I_{corr}
0	0.0344	0.008	1.05	19.3	20.27
0.74	0.0368	0.016	1.063	18.53	19.70
0.95	0.0428	0.026	1.083	17.65	19.12
2.33	0.0464	0.030	1.092	16.87	18.43
3.17	0.052	0.040	1.11	15.688	17.42
3.97	0.0572	0.047	1.127	15.30	17.24
4.45	0.0604	0.052	1.138	15.03	17.10
5.18	0.0648	0.059	1.153	14.54	16.77
6.14	0.0724	0.067	1.174	13.96	16.39
6.56	0.076	0.072	1.186	13.37	15.86
6.98	0.080	0.078	1.200	12.78	15.34
7.88	0.084	0.084	1.213	12.49	15.15
8.68	0.0892	0.092	1.232	11.49	14.16
9.42	0.0936	0.099	1.248	11.51	14.37

One can notice that after correction for the inner filter effect, one can still observe a decrease in the fluorescence intensities of the Trp residues. This decrease is a result of the TNS binding to the flavodehydrogenase.

Experiment 2

Correction for the volume

[TNS] (μM)	Volume in the cuvette (μl)	I_{rec}	I_{vol}
0	1100	17.4	17.40
2	1110	17.1	17.26
4	1120	16.6	16.90
6	1130	16.4	16.85
8	1140	16.4	17.0

Correction for the inner filter effect

[TNS] (μM)	$\lambda_{ex} = 280$ nm $OD_{(experiment)}$	$\lambda_{em} = 355$ nm $OD_{(experiment)}$	f_c	I_{vol}	I_{corr}
0	0.0360	0.012	1.057	17.40	18.40
2	0.0480	0.034	1.099	17.26	18.96
4	0.0588	0.045	1.127	16.90	19.00
6	0.0740	0.058	1.164	16.85	19.61
8	0.0856	0.070	1.196	17.00	20.33

One can see that after correction for the dilution and the inner filter effect, the measured intensities in the presence of TNS reach that measured for the protein alone. This indicates that binding of TNS to the flavodehydrogenase does not occur, and the protein has lost its specific binding site for the TNS.

This experiment can be repeated with other proteins such as bovine serum albumin, a protein that is commercially very easily available.

17.2.4.12   *Exercise 12*

Plotting the velocity expressed as optical density per minute vs. substrate concentrations in the absence and presence of the three inhibitors yields the results shown in Figure 17.5.

The curves shown indicate that ONPTG, maltose, and melibiose, in the range of the concentrations used, inhibit the O-nitrophenyl galactoside hydrolysis rate. We can see from the curves that a plateau is not reached at the high concentrations of inhibitors, meaning that the maximum velocity V_M is not reached. Thus, in order to determine V_M and K_M, reciprocal plots should be drawn (Figure 17.6).

ONPTG acts as a competitive inhibitor, maltose as a noncompetitive inhibitor, and melibiose as a noncompetitive inhibitor.

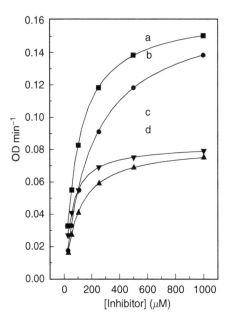

Figure 17.5   Velocity of hydrolysis of *O*-nitrophenyl galactoside with β-galactosidase in the absence (a) and presence of inhibitors, 3×10^{-4} M ONPTG (b), 0.26 M maltose (c), and 0.17 M melibiose (d).

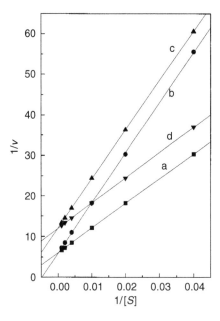

Figure 17.6   Determination of K_M and V_M values of *O*-nitrophenyl galactoside with β-galactosidase in the absence (a) and presence of inhibitors, 3×10^{-4} M ONPTG (b), 0.26 M maltose (c), and 0.17 M melibiose (d). The values found are: (a) $K_m = 100$ μM and $V_{max} = 0.165$ OD min^{-1}. (b) $K_m = 208$ μM and $V_{max} = 0.169$ OD min^{-1}. (c) $K_m = 100$ μM and $V_{max} = 0.082$ OD min^{-1}. (d) $K_m = 5.92$ μM and $V_{max} = 0.083$ OD min^{-1}.

Index

www.ingramcontent.com/pod-product-compliance
Lightning Source LLC
Chambersburg PA
CBHW080636260225
22555CB00012B/151